HRW
ALGEBRA ONE
INTERACTIONS
Course 2
COOPERATIVE-LEARNING ACTIVITIES

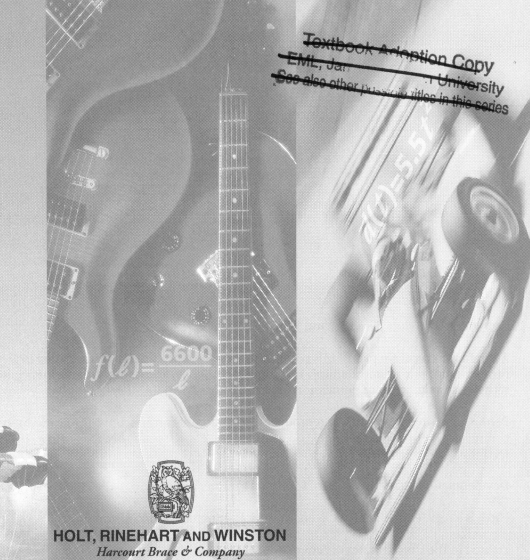

$$f(\ell) = \frac{6600}{\ell}$$

HOLT, RINEHART AND WINSTON
Harcourt Brace & Company
Austin • New York • Orlando • Atlanta • San Francisco • Boston • Dallas • Toronto • London

To the Teacher

HRW Algebra One Interactions Course 2 Cooperative-Learning Activities contain one-page blackline masters for each of the 66 lessons in *HRW Algebra One Interactions Course 2*. These masters provide structured activities for students to perform in small groups or in pairs to reinforce the mathematical content in the lessons. Directions designate specific roles or responsibilities that facilitate student cooperation and participation.

Developmental assistance by B&B Communications West, Inc.

HRW is a registered trademark licensed to Holt, Rinehart and Winston, Inc.

Printed in the United States of America

ISBN 0-03-051308-1

1 2 3 4 5 6 7 066 00 99 98 97

TABLE OF CONTENTS

Cooperative-Learning Activity
1.1 Tic-Tac-Toe

Group members: 2

Materials: unruled paper, pencil, and index cards

Responsibilities: Write and solve equations in order to play a modified version of tic-tac-toe.

Preparation: Each player should copy five of the equations listed below onto index cards and write five additional equations on his or her own. All of the equations should be solved.

Procedures: **1.** Use the game board below.

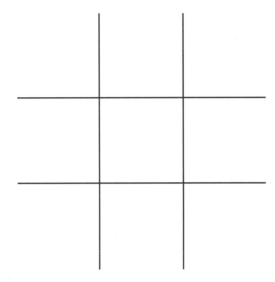

2. Decide which player will be X and which will be O.

3. Select one of your opponent's equation cards without looking at the equation. Solve the equation. Have your partner check your answer. If it is correct, place an X (or O) on the game board. If your answer is incorrect, you lose a turn.

4. Continue to play until one player has three Xs or three Os in a row.

Equations

Player 1	Player 2
$x - 16 = -43$	$y - 19 = -54$
$y + 3.2 = 97$	$x + 5.3 = 81$
$-6.4 + x = 25$	$-12.9 + y = 32$
$y + 37 = -24$	$y + 48 = -61$
$x - 108 = -205$	$x - 99 = -189$

Cooperative-Learning Activity
1.2 Architect for a Day

Group members: 2–4

Materials: graph paper, pencil, and a ruler

Responsibilities: Make a scale drawing of a room.

Preparation: Find a room in your school that is accessible for your group to take actual measurements. You may choose the library, gymnasium, cafeteria, or a classroom.

Procedures:

1. As a group, decide what scale you will use before you begin making your scale drawing. For example:

$$\frac{\text{Actual measurement}}{\text{Scale measurement}} = \frac{2\ \text{feet}}{1\ \text{inch}}$$

Consider the size of the paper used for your drawing before you come to a decision about the scale.

2. Use a ruler, tape measure, or meter stick to measure the length and width of the room. Have two members of the group hold the measuring device while the others check the measurements and record the information on a piece of paper.

3. Write and solve proportions to find out what your scale drawing should look like. For example, if the width of the room is 20 feet, set up the following proportion:

2 feet : 1 inch = 20 feet : x inches

Solve by using cross multiplication.

$$\frac{2}{1} = \frac{20}{x}$$
$$2x = 20$$
$$x = 10\ \text{inches}$$

Therefore, the width of the room in your scale drawing should be 10 inches.

4. Write a summary of how knowing actual measurements and scale measurements can help in certain occupations. Compare your summaries with other groups in your class.

_____

_____

_____

_____

Cooperative-Learning Activity
1.3 Forensic Science

Group members: 2–3

Materials: human skeletal bone chart, paper, pencil, and a meter stick

Responsibilities: Calculate the height of a person by using a formula.

Preparation: Obtain a diagram of a skeleton from a science book or encyclopedia that outlines the bones of the body. Specifically, you need to know where the femur, tibia, humerus, and radius bones are located.

Procedures: Forensic scientists use formulas to find the exact physical dimensions of crime victims in order to positively identify them. If a person's body is found after it has already begun to decompose, the scientist can apply certain formulas to find out information about the victim when he or she was alive. For example, they use the lengths of the femur (F), the tibia (T), the humerus (H), or the radius (R) to calculate the height of the person (h). When the length of one bone is known, one of four formulas is used to determine the person's height. Different formulas are used for males and females. All measurements are taken in centimeters.

1. Your group has been asked to test the accuracy of these formulas.

2. Take the measurements listed in the table for all of the members in your group.

Member	Height (cm)	Femur (cm)	Tibia (cm)	Humerus (cm)	Radius (cm)

3. Apply the formulas.

FORMULAS

Male	Female
$h = 69.089 + 2.238F$	$h = 61.412 + 2.317F$
$h = 81.688 + 2.392T$	$h = 72.572 + 2.533T$
$h = 73.570 + 2.970H$	$h = 64.977 + 3.144H$
$h = 80.405 + 3.650R$	$h = 73.502 + 3.876R$

4. Do the lengths of your bones accurately predict your actual height?

_____

_____

Cooperative-Learning Activity
1.4 Algebraic Four-in-a-Row

Group members: 3–5

Materials: paper, pencil, paper cup or bag, and a colored pencil for each student

Responsibilities: Make a game board and solve equations correctly to be the first player to get 4 squares in a row.

Preparation: Each player must make a 4 × 4 game board using the numbers given in the example below. The numbers may be written in any order on the game board. All of the equations given should be written on a piece of paper and folded in half. Each student should place the slips of paper into his or her paper cup or bag.

Procedures: 1. The first player chooses a slip of paper from the cup and shows everyone in the group which equation is written on it. Each player solves the equation and finds the solution on the board. As a group, verify the solution before a mark is put on the game board. If the answer is correct, mark an X with the colored pencil over the square. Players who answer incorrectly do not mark a square on their game board.

Solutions To Be Written on the Game Board

-7	0	-3	3
-5	8	$\frac{13}{8}$	$\frac{27}{4}$
$-\frac{2}{3}$	$\frac{3}{2}$	$\frac{7}{2}$	-30
$-\frac{1}{2}$	42	2	-16

2. Take turns choosing the equation. The first player to get four Xs in a row, column, or diagonal wins.

Equations

$3 - 4y = 10y + 10$ 　　 $\frac{2}{3}n + 8 = \frac{1}{3}n - 2$ 　　 $\frac{3}{4}n + 16 = 2 - \frac{1}{8}n$

$6y + 2 - 4 = -10$ 　　 $5 - \frac{1}{2}(b - 6) = 4$ 　　 $4(2x - 1) = -10(x - 5)$

$4(2a - 8) = \frac{1}{7}(49a + 70)$ 　　 $2(x - 3) + 5 = 3(x - 1)$ 　　 $2(x + 3(x - 1)) = 18$

$4x - 9 = 7x + 12$ 　　 $3w + 2 = 2 - 10w$ 　　 $8n - 13 = 13 - 8n$

$8x - 15 = 4x + 12$ 　　 $\frac{1}{3}s + \frac{3}{4} = \frac{5}{6}s - 1$ 　　 $8p - 5(p + 3) = (7p - 1)3$

$-5(2 - 3t) = 7 - 2(t - 3)$

Cooperative-Learning Activity
1.5 The Missing Number

Group members: 2–4

Materials: paper, index cards, and a pencil

Responsibilities: Find the value of a variable to make a set of inequalities true.

Preparation: Write each inequality on an index card.

Procedures:
1. Shuffle the cards and give an equal number of cards to each member of the group.

2. There is one solution that makes all of the inequalities true. One player begins by turning over one of his or her cards, makes up a clue about the inequality, and tells it to the group. For example, "The unknown number plus -3 is less than or equal to 10." Group members can translate the verbal inequality into symbols and make notes about a possible solution.

3. Group members take turns giving clues about each of the inequalities. Play continues until one member knows the value of x. Use the workspace provided below each inequality to substitute this number into the inequalities and see if it holds true for all of the inequalities given.

4. When the value of x is discovered, the group member who discovered it should go immediately to the teacher to confirm the value.

Inequalities

$10 \geq -3 + x$ $\qquad$ $5x + 4 < 6x$ $\qquad$ $-12x \leq 30$

$-x \leq 44$ $\qquad$ $6x + 4 \geq 5x + 4$ $\qquad$ $16x < 96$

$-8x \geq -72$ $\qquad$ $2.4x + 13 \leq 5x$ $\qquad$ $4x - 6 > 6x - 20$

$6 - 11x \leq -3$ $\qquad$ $\dfrac{x}{6} - 16 < -9$

Cooperative-Learning Activity
1.6 Word Wise

Group members: 2 or 4

Materials: pencil

Responsibilities: Translate statements into algebraic equations or inequalities.

Procedures:
1. One group member is chosen to read the statements below to the group. As each statement is read, all group members should translate the statements into algebraic equations or inequalities and write them in the space provided.

2. After each equation or inequality is written, members should solve the problem. Check each member's calculations. Award two points if the member wrote and solved the problem correctly. Award one point if the translation is correct but the solution is not.

3. Play continues until one member scores 10 points.

4. At the conclusion of this round, challenge members to write their own statements. Play the game again.

Equations and Inequalities

The distance between a number and −1 is less than 4. _____

The distance between a number and −7 is less than 2. _____

The distance between a number and 8 is greater than or equal to 1.

_____

The distance between a number and −5 is equal to −3. _____

The distance from 0 to x is 4 units. _____

The distance between a number and −3 is less than 1. _____

The distance between a number and 2 is greater than or equal to 1.

_____

The distance between 3 and a number is greater than 0.

_____

Cooperative-Learning Activity
2.1 Rise, Run, and Slope

Group members: 3

Materials: 2 meter sticks, 2 levels, tape, paper, and a pencil

Roles: **Horizontal Holder** holds meter stick horizontally and levels it

Vertical Holder holds meter stick vertically and levels it

Taper/Recorder tapes levels in place and reads and records measurements

Preparation: Your teacher will direct your group to one or more objects that have a slope, or an incline. Your group will measure the rise and run of the incline and determine the slope.

Procedures:

1. To find the rise, or vertical distance from the ground, the Horizontal Holder places the 0-cm end of one meter stick at a high point on the object and holds it horizontally.

2. The Horizontal Holder levels the meter stick by placing one level on the horizontal meter stick, then raises or lowers the 100-cm end until the bubble rests between the two marks on the level. The Taper/Recorder tapes the meter stick to hold it in place.

3. The Vertical Holder aligns the other meter stick so that it crosses the horizontal stick and the 0-cm end rests on the object. The Vertical Holder levels the meter stick vertically by moving the 100-cm end left or right until the bubble rests between the two marks on the level. The Taper/Recorder tapes the meter stick to hold it in place, then reads the measurements where the two meter sticks intersect and records them in the table below.

Objects	Sketch	Rise	Run	$\frac{Rise}{Run}$ = Slope

4. As a group, draw a sketch of the object or objects you measured, and then calculate the slope.

5. As a class, discuss the meaning of the different slopes that the groups calculated.

Cooperative-Learning Activity
2.2 Finding the Most Cost-Efficient Appliance

Group members: 4

Materials: 4 different Energy Guide labels for air conditioners or other appliances, graph paper, graphics calculator, paper, and a pencil

Responsibilities: Draw the graph of a linear function and find the line of best fit.

Preparation: An Energy Guide label provides information about the energy costs for running an appliance such as an air conditioner. The energy efficient rating (EER) is used to determine how much it will cost you to run the model yearly.

ENERGYGUIDE
↓

Yearly hours of use		250	750	1000	2000	3000
Estimated yearly $ cost shown below						
Cost per kilowatt hour	2¢	$3	$9	$12	$24	$37
	4¢	$6	$18	$24	$49	$73
	6¢	$9	$27	$37	$73	$110
	8¢	$12	$37	$49	$98	$146
	10¢	$15	$46	$61	$122	$183
	12¢	$18	$55	$73	$146	$220

Procedures: Use the following procedures to determine which of several appliances is the most cost-efficient.

1. Find your local energy rate and record it here:

   Cost per kilowatt hour: _____

2. Each group member selects a different Energy Guide label and sets up a graph with the yearly hours of use on the horizontal axis and the estimated cost on the vertical axis.

3. Each group member chooses the cost per kilowatt hour closest to your local energy rate and uses the information on the label to form a set of ordered pairs relating yearly hours of use to yearly costs.

4. Each member displays his or her data as a scatter plot and connects the points. As a group, explain why the data can be modeled by a linear function.

5. Each member uses a graphics calculator to find a line of best fit for the hours of use and yearly cost. Then use the line of best fit to estimate the total cost of using the appliance for 500 hours per year.

6. As a group, compare the graphs to determine which appliance would cost the least to run for 500 hours per year.

7. Based on your results, decide as a group whether you would buy an appliance based only on its initial price or whether you would also consider the energy costs per year.

Cooperative-Learning Activity
2.3 Number of Intersection Points

Group members: 2

Materials: 5 × 5 geoboard, rubber bands, graphics calculator, and a pencil

Responsibilities: Model systems of linear equations and determine their common solution.

Preparation: The graph of a linear equation is a straight line. In a system of linear equations, each pair of lines may intersect at a point that represents their common solution.

Procedures:

1. Use the geoboard to model a coordinate plane. Let the peg at the lower left-hand corner of the geoboard represent the origin. Stretch a rubber band between the pegs representing the points (0, 1) and (4,0). Next, stretch a rubber band between the pegs representing the points (0, 2) and (3, 0).

2. Count the number of times that the rubber bands intersect, and record this number in the chart below.

Number of rubber bands	Number of intersections
1	
2	

3. Take turns with your partner. Repeat the process of stretching rubber bands and counting intersection points until all of the pegs are connected. All stretches must begin on one of the borders and end on the other border.

4. Study the data points. What appears to be the relationship between the number of rubber bands and the number of times they intersect?

_____

5. Choose any two rubber bands on the geoboard and locate the point of intersection. With your partner, make a reasonable estimate to approximate the common solution. Using the pegs as ordered pairs of points, write an equation to model each of the two rubber bands. Graph your equations on a graphics calculator. Then find the common solution.

6. Describe how well your estimate matches the results from the graphics calculator.

Cooperative-Learning Activity
2.4 Finding the Number of Students

Group members: 2

Materials: different colored counters, and a pencil

Responsibilities: Model two equations and manipulate counters to find their solution.

Preparation: The number of students in a certain algebra class never changes. Each day the students come to class, and they always sit in the same seats, either on the right or the left side of the teacher's desk.

One day Miquelina, who always sits on the left side, decides to sit on the right side of the room. With this change, there is now an equal number of students sitting on each side of the teacher's desk. The next day, Miquelina moves back to where she always sits. But then Ricardo, who always sits on the right side of the desk, moves over to the other side. With this move, there are now twice as many students sitting on the left side as on the right side.

Procedures: Use the following procedure to find the number of students who sat on the right side and left side of the teacher's desk before Miquelina moved.

1. To visually represent the changing relationship between the number of students on the right and left sides of the teacher's desk, use different colored counters. Work with your partner to manipulate your counters according to the chart below, making sure that each condition of the problem is satisfied.

Word description	Left side	Right side	Equation
Possible number of students			
One student moves from left to right side; equal number of students on each side			
Student moves back to left side; another student moves to left side			

2. How many students originally sat on the right and the left sides of the teacher's desk? _____

Cooperative-Learning Activity
2.5 Find the Common Solution

Group members: 4

Materials: graph paper, calculator, pencil, and paper

Responsibilities: Solve and graph systems of equations.

Preparation: Use the procedure below to find out what unusual property the systems of equations shown have in common.

Procedures:
1. Each member solves one of the following system of equations.

   a. $\begin{cases} x + 2y = 3 \\ 3x + 4y = 5 \end{cases}$

   _____

   b. $\begin{cases} 89x + 90y = 91 \\ 6x + 7y = 8 \end{cases}$

   _____

   c. $\begin{cases} -3x - 2y = -1 \\ 4x + 5y = 6 \end{cases}$

   _____

   d. $\begin{cases} 4x + 5y = 6 \\ -5x - 6y = -7 \end{cases}$

   _____

2. As a group, compare your solutions and discuss what you notice about them.

3. Notice the coefficient of the x-terms, the coefficient of the y-terms, and the constant terms for each system of equations. As a group, describe the pattern formed by these numbers.

4. Each member graphs one of the following equations on the same piece of graph paper and the same coordinate system:

$$\begin{cases} x + 2y = 3 \\ 3x + 4y = 5 \\ -3x - 2y = -1 \\ 4x + 5y = 6 \end{cases}$$

5. Describe what you notice about the points of intersection of the graphs.

6. Work in pairs to write two systems of equations with the same solution as the systems in this activity. Then check your solution by using the elimination method.

Cooperative-Learning Activity
2.6 Disappearing Lines

Group members: 2 or 3

Materials: graph paper, scissors, ruler, and a pencil

Responsibilities: Copy and graph systems of parallel lines.

Preparation: Segments of the graphs of $x = 2$, $x = 4$, and $x = 6$ are drawn on the same grid, along with the graph of the diagonal line $y = x$. This line intersects $x = 2$ at the point $(2, 2)$ and intersects $x = 4$ at $(4, 4)$ and intersects $x = 6$ at $(6, 6)$.

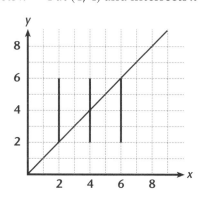

Procedures:

1. One group member copies and cuts out the grid and then cuts along the line $y = x$.

2. Another member slides the top half of the grid to the right until the segment on $x = 4$ coincides with the segment on $x = 2$, and the segment on $x = 6$ coincides with the segment on $x = 4$.

3. As a group, count the number of line segments. Describe what happened.

4. Another member slides the top half back. Describe what happens.

5. As a group, graph segments of $x = 2$, $x = 4$, $x = 6$, $x = 8$, $x = 10$, $x = 12$, $x = 14$, $x = 16$, $x = 18$, and $x = 20$ on the same grid. Then draw the graph of $y = x$ so that it intersects each vertical line at one point.

6. One member cuts out the grid and then cuts along the line $y = x$. Another member slides the bottom half to the left until the line segments coincide.

7. As a group, count the number of line segments. _____

8. As a group, explain how the puzzle works.

_____

_____

Cooperative-Learning Activity
2.7 Treasure Hunt

Group members : 3

Materials: different colored pencils, graph paper, and a ruler

Responsibilities: Graph a system of linear inequalities and locate a treasure.

Preparation: Each group will use linear functions and systems to hunt for buried treasure. You must find all three treasures, not just one. You have found an old map that has three inequalities written on it, each representing one of three roads, A, B, and C, and the area bounded by them.

Road A	$2x + 3y \geq 6$
Road B	$x - y \leq -2$
Road C	$6x - y \leq 18$

Procedures:

1. Take turns graphing roads A, B, and C on the same grid. As a group, locate and shade the area bounded by the roads.

2. To narrow down the location, one member writes the equation of a line that intersects roads A and B and passes through the shaded region. A second member writes that equation of a line that intersects roads B and C and passes through the shaded region.

_____

_____

3. The third member writes an equation that intersects roads A and C and passes through the shaded region.

_____

4. As a group, write three equations that intersect at one point within this shaded region. That point will be where the treasure is located.

_____

_____

_____

Cooperative-Learning Activity
2.8 Reversing Digits

Group member: 4

Materials: calculator, and a pencil

Responsibilities: Create a chart to determine some effects of reversing the digits of a 3-digit number.

Procedures:

1. Each member chooses a 3-digit number composed of three unique digits and records this number in the chart below.

Number	Number with digits reversed	Difference	Difference with digits reversed	Sum

2. Each member then reverses the digits of the 3-digit number and records the results in the chart.

3. Next, each member finds the difference between the original number and the number with digits reversed by subtracting the smaller number from the larger number. The result is recorded in the "Difference" column on the chart.

4. Now each member reverses the digits of the number that represents the difference and records the results.

5. Then each member finds the sum of the number representing the difference and the number representing the difference with digits reversed. This sum is recorded in the "Sum" column on the chart.

6. As a group, describe what you notice about the sum for each number.

_____

7. As a group, explain why this puzzle works.

_____

Cooperative-Learning Activity
3.1 Matrix Make-up

Group members: 2–4

Materials: maps, atlas, and a pencil

Responsibilities: Groups will arrange statistical information in matrices of different sizes.

Preparation: Maps, an atlas, and other reference material should be available for each group at the start of the activity.

Procedures: 1. Think about 5 popular cities located in your state. Using your resource materials, find the distances between 5 different cities. Arrange this information in a 5×5 matrix.

_____ _____ _____ _____ _____

_____ _____ _____ _____ _____

_____ _____ _____ _____ _____

_____ _____ _____ _____ _____

_____ _____ _____ _____ _____

2. As a group, decide how this information can be used and why the matrix format is useful.

_____

_____

3. As a group, choose three states from a map of the United States. Research and find out at least three statistics about each state. Arrange this information in a 3×3 matrix.

_____ _____ _____

_____ _____ _____

_____ _____ _____

Cooperative-Learning Activity
3.2 Matrix Madness

Group members: 2–5

Materials: pencil, 50 index cards, number cube, and a coin

Roles: **Leader** shuffles cards and asks the teacher for assistance, if needed

Recorder writes numbers on index cards and records group's matrices

Roller rolls a number cube to determine matrix dimensions

Drawer draws index cards to fill the matrix dimensions

Flipper flips a coin to determine procedure performed on the matrices

Preparation: As a group, you will generate numbers to form matrices and use the matrices to add or subtract. The Recorder writes each of the following numbers on an index card:

$$-1 \text{ through } -20$$
$$0 \text{ through } 20$$
$$0.1 \text{ through } 0.9$$

Procedures: 1. The Leader shuffles the cards and places them facedown on the desk. The Roller rolls the number cube. The first number rolled is the number of rows that the group's matrix will have. The Roller rolls the number cube again. The second number rolled is the number of columns that the group's matrix will have. For example, if a 3 and a 2 are rolled, your group will be making a 3×2 matrix.

2. To complete the group's matrix, the Drawer draws the number of cards that he or she will need to fill each entry. For example, a 3×2 matrix has 6 entries, so the Drawer would draw 6 cards from the deck. The Recorder records the group's matrix on a separate piece of paper.

3. Repeat Procedures 1 and 2. The group should now have two matrices.

4. Then the Flipper flips the coin. If the result is heads, then the group will add the matrices. If the result is tails, then the group will subtract the matrices. The Recorder records the group's work in the space provided.

5. Repeat Procedures 1–4 until each group member has had a chance to roll the number cube at least once. You may reuse the cards, if necessary.

6. Have the group members take turns checking all computations with a calculator.

Cooperative-Learning Activity
3.3 Candy Products

Group members: 4–6

Materials: pencil and a small package containing at least two types of assorted candies

Responsibilities: As a group, you will display data about the frequency of different colored candies and different types of candy.

Preparation: Frequency is the number of times that something happens or is chosen.

Procedures:

1. Each group member counts the number of each color of candy and the type of candy in their package. Record the data in a chart similar to the one provided.

Group member: Tim _____

Color	Brand X	Brand Y
Red	ЖШ ‖	‖‖
Brown	0	0
Yellow	ЖШ	‖
Blue	‖‖‖	0
Green	‖	ЖШ ‖
Orange	‖	‖‖
Other	ЖШ ‖	0

2. Use the group's data to create a matrix of at least 3 colors of candy. Then create a matrix that represents the types of candy and the total number of each color from the entire group's supply. As you construct your matrices, keep in mind that they will be multiplied by one another. Your matrices should resemble the ones illustrated below.

$$
\begin{array}{c}
 \\
\text{Tim} \\
\text{Stacey} \\
\text{Alan} \\
\text{Kim}
\end{array}
\begin{array}{ccc}
\text{Red} & \text{Blue} & \text{Green}
\end{array}
\left[
\begin{array}{ccc}
7 & 4 & 2 \\
5 & 3 & 8 \\
8 & 2 & 0 \\
4 & 9 & 2
\end{array}
\right]
\qquad
\begin{array}{c}
 \\
\text{Red} \\
\text{Blue} \\
\text{Green}
\end{array}
\begin{array}{cc}
\text{Brand X} & \text{Brand Y}
\end{array}
\left[
\begin{array}{cc}
16 & 8 \\
6 & 12 \\
6 & 6
\end{array}
\right]
$$

3. After the matrices are constructed, multiply them together in the space provided.

4. As a group, discuss how showing information in a matrix might be helpful to a person who works for the candy company.

Cooperative-Learning Activity
3.4 Matrix Inverse

Group members: 3

Materials: pencil

Responsibilities: Form a matrix of the number of hours spent doing homework and find the inverse matrix.

Preparation: Find the inverse of a matrix.

Procedures: **1.** Each member of the group will complete one of the following matrices with the number of hours that they spend doing homework each day, from Monday through Thursday.

    a. Monday   Tuesday   Inverse matrix

          _____   _____

          _____   _____

          Wednesday   Thursday

    b. Monday   Tuesday   Inverse matrix

          _____   _____

          _____   _____

          Wednesday   Thursday

    c. Monday   Tuesday   Inverse matrix

          _____   _____

          _____   _____

          Wednesday   Thursday

2. With your group, trade your matrices and find the inverse matrix for the matrix you receive. Record your results in the space provided.

3. As a group, discuss how you can use your inverse matrix to prove that it is an inverse matrix.

Cooperative-Learning Activity
3.5 Ideal Weight

Group members: 2–3

Materials: paper, and a pencil

Responsibilities: The group uses formulas to solve systems of equations and writes a generalization about the patterns they notice.

Preparation: You can calculate your "ideal" weight from information about your bone structure. For example, a female with average-size bones should weigh 100 pounds for the first 5 feet of height and an additional 5 pounds for each inch over 5 feet. A male with average-size bones should weigh 106 pounds for the first 5 feet of height and 6 additional pounds for each inch over 5 feet.

Procedures:
1. Each member solves one of the following systems of equations. As a group, discuss the results.

   a. The formulas used to find the ideal weight, y, of a person with average-size bones based on his or her height in inches, x, are as follows:

   $$\text{Male:} \quad y = 6x - 254$$
   $$\text{Female:} \quad y = 5x - 200$$

   Solve the system of equation by using matrices.

   b. The formulas used to find the ideal weight of a person with large bones are as follows:

   $$\text{Male:} \quad y = 6.6x - 279.4$$
   $$\text{Female:} \quad y = 5.5x - 220$$

   Solve the system of equation by using matrices.

   c. The formulas used to find the ideal weight of a person with small bones are as follows:

   $$\text{Male:} \quad y = 5.4x - 228.6$$
   $$\text{Female:} \quad y = 4.5x - 180$$

   Solve the system of equation by using matrices.

2. As a group, write a generalization about the pattern(s) your group noticed.

   _____

   _____

   _____

   _____

Cooperative-Learning Activity
4.1 Checkerboard Squares

Group members: 3

Materials: one checkerboard, dime, nickel, quarter, tape, and a ruler

Responsibilities: Each member of the group will explore the influence of probability in coin tossing games by tossing a coin a preset number of times onto a checkerboard from a predetermined distance.

Preparation: The object is to make the coin land inside a red or black square, not in contact with any of the lines on the board. A successful coin toss gains one point. An unsuccessful toss loses one point. If the coin does not land on the checkerboard, the toss is not counted. The winner is the player who scores the greatest number of points after each player has had a turn.

Procedures:
1. Place a checkerboard on the desk. Then make a line with tape a measured distance away from the board. The initial distance should be no less than 3 feet away.

2. Each player will stand behind the line and toss a coin, either a dime, nickel, or quarter onto the board. All players will use the same coin for each round and will make the same number of tosses. Tosses that do not land on the board will not count. Each player will make a minimum of 50 tosses per round.

3. Predict each player's score. Set up a chart to record the predicted and actual scores of each player.

4. Play additional rounds but change the conditions. For example, use a different coin or increase the distance from the board. Then change the number of tosses that each player takes.

5. Does the change in conditions seem to affect a player's performance or increase the accuracy of the predicted performance? Does practice seem to improve the overall performance of all players? Can you predict which players will get a higher score than others?

   _____

   _____

6. As a group, describe how the element of chance is responsible for the scores that were achieved.

   _____

   _____

Cooperative-Learning Activity
4.2 Exploring Simulation

Group members: 3

Materials: paper, pencil, and a coin

Roles: Leader tosses a coin and counts frequencies

Recorder records the answers for the group from the coin tosses

Reporter summarizes the group's work and writes a summary

Preparation: Each team will toss a coin to simulate guessing on a 10-question true-false quiz. A trial consists of tossing a coin 10 times, one for each question on the quiz. A successful trial occurs when 7 or more correct answers are obtained.

Procedures:

1. The members choose their roles. To randomly determine the answers to 10 questions on a true-false quiz, the teacher tosses a coin 10 times. Heads represents a false answer, and tails represents a true answer. The teacher uses the results to write an answer key.

2. Before tossing the coin, the Leader estimates the group's chances of getting 7

   or more of the 10 questions correct by guessing. _____

3. The Leader tosses the coin. If heads comes up, the Recorder writes an *F* for the question. If tails comes up, the Recorder writes a *T* for the question. Then the Leader tosses the coin 9 more times, and the Recorder writes the results in a table that has 10 rows, representing the result of the coin toss, and 20 columns, representing the trial number.

4. Now use the teacher's answer key to determine the correct answers. The Leader counts the number of correct answers and the Recorder writes the results. Repeat this trial 19 more times and record the results. Rotate roles after each trial.

5. The Leader counts the number of times that 7 or more questions were answered correctly and the Recorder records the results. Find the experimental probability of getting 7 or more questions correct.

6. Each member compares the group's estimate of guessing correctly with the experimental probability. Are the results the same?

   _____

7. Combine all of the data from your group. The group discusses how to find the entire team probability of guessing 7 or more questions correct. The Reporter writes a summary of the discussion.

   _____

   _____

Cooperative-Learning Activity
4.3 Extending Statistics

Group members: 2

Materials: graphics calculator or a number cube

Responsibilities: The group will conduct an experiment to test whether the results of a game are predictable.

Preparation: Frequency is the number of times an even takes place.

Procedures:

1. Use the table below to keep a record of the results.

Number	Frequency

2. Work in pairs. Before the experiment begins, each member predicts the number of points that he or she expects to score in the game.

3. Roll a number cube and record the number that comes up on the cube in the number column. If a 1, 2, 4, 5, or 6 is rolled, the member earns that many points. A roll of 3 loses three points. A graphics calculator may also be used to simulate the game.

4. Each player calculates the total number of points from all rolls. The player with the most points wins the game. However, the game ends when the number 3 appears on 10 rolls of the number cube.

5. Calculate your score by multiplying the frequency by the number rolled for each number and record your answer. Did you overestimate or underestimate your score?

6. Repeat the game nine more times and record your score for each game. Combine all the data from your class. Use the data from the entire class to construct a bar graph.

7. Calculate the range, mean, median, and mode for the class data. Then discuss any patterns you notice. Is this game random or is it predictable?

8. If a roll of 1 loses one point and a 2, 3, 4, 5, or 6 earn that many points, would the average score be the same, higher, or lower than average score of this game? Discuss.

Cooperative-Learning Activity
4.4 Exploring the Addition Principle of Counting

Group members: 3

Materials: 4 coins of the same type and a square sheet of paper

Roles: **Leader** asks teacher for assistance, if needed

Recorder records answers for the group

Reporter reports the group's work to the class

Preparation: Each player arranges the coins in specific heads-and-tails combinations in the fewest possible moves.

Procedures: 1. The members choose their roles. The Leader places four coins on the desk, tails up.

2. To arrange the coins so that the four coins show heads, the Leader must turn over three coins at a time. Repeat this process until all coins show heads. The player with the least number of moves wins the game.

3. The group members play the game, and the Recorder records the group's moves in a table like the one below that extends to the eighth move.

Heads and Tails

	First coin	Second coin	Third coin	Fourth coin
First move				
Second move				
Third move				

4. Examine your moves. How many different moves did you make? Does your first move affect the moves that follow? Compare each group member's moves with the other group members. Which player had the least number of moves?

5. Examine your first move. How many different ways can you make the first move? Notice the position of the coin that was not turned over initially.

6. Rotate roles. Play the game again and the Recorder records the group's moves. Did you make more or less moves than the previous game? Did you win this round? Now examine your second move. How many different ways can you make the second move? Does the second move depend on the first move?

7. Continue playing the game. The winner is the first player who can turn over the coins, showing all tails or heads, in four moves. The Reporter reports the group's results to the class.

Cooperative-Learning Activity
4.5 Multiplication Principle of Counting

Group members: 2

Materials: colored pencils

Roles: Recorder records the total number of different paths

Reporter summarizes the group's work and steps used to find the results

Preparation: As a group, you will solve a puzzle that requires determining all of the different ways that an event can occur. Each group will explore counting methods in which choices are made one after the other.

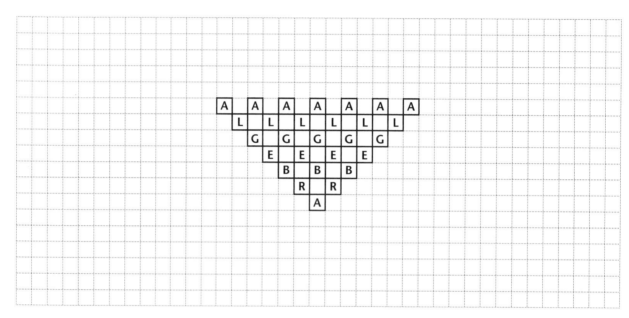

Procedures: 1. The members choose their roles. Each member examines the diagram. To solve this puzzle, find the total number of different paths that spell out the word *algebra*, following these rules:

- Start with the letter *a* on the first row.
- You can move down, to the left, or to the right.
- You must move one letter at a time.
- You may never pass a letter without using it.

Hint: Use different colored pencils to draw the paths.

2. The Recorder records the total number of different paths, and the Reporter summarizes the steps used to find the results.

_____

_____

_____

Cooperative-Learning Activity
4.6 Theoretical Probability

Group members: 2

Materials: 2 index cards, 10 markers, paper, and a pencil

Roles: **Recorder** records answers for the group

Reporter summarizes the group's work

Preparation: A game is considered fair when each player's probability of winning is equally likely. In this activity, you will determine which of the two games is fair.

Procedures: 1. Label one index card *ODD*. Label the other index card *EVEN*. Label each marker with the numbers 0 through 9.

Game A

1. The members choose their roles. One member chooses an index card and gives the other card to his or her partner.
2. Each player randomly selects one of the 10 markers. The first marker is replaced before the second player draws. The following condition determines the winner: **If the sum of the two numbers drawn is even, the player with the *EVEN* index card wins. If the sum is odd, the player with the *ODD* index card wins.**
3. Play the game several times. The Recorder records the results in a table that has two rows labeled "Even" and "Odd" and one column labeled "Total wins."
4. Each player estimates his or her chances of winning, and the Reporter summarizes the group's results.

Game B

This game uses the same rules as game A, but the 0 marker is removed. Thus, you can only choose from markers labeled 1 through 9. Play the game several times and record the results. Estimate your chances of winning game B.

2. How many ways can *EVEN* or *ODD* win game B? How many outcomes have an even sum?

_____

3. Find the probability that *EVEN* will win. Then find the probability that *ODD* will win. Compare the theoretical probability of winning game B with your estimated chances of winning.

_____

4. Determine if game B is fair. Is game A fair? Explain.

_____

Cooperative-Learning Activity
4.7 Independent Events

Group members: 2

Materials: colored pencils

Preparation: Determining probabilities in a complex situation can be done by examining a similar situation.

Procedures: This diagram shows Sam's house (*S*) in relation to Harry's house (*H*). Sam plans to walk from his house to Harry's house. Sam decides to walk either east or north. How many ways can he reach Harry's house?

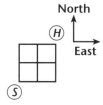

To keep track of the possible paths, write a letter at each intersection to name the ways that Sam can get from his house to Harry's house by traveling only north or east.

1. How many ways are there from *S* to *D*? _____

2. How many ways are there from *S* to *H*? _____

Suppose that Sam decided that at each intersection he would flip a coin to determine whether he would go north (heads) or east (tails). Look at a simpler case. If Sam starts at *S*, then the probability that he goes to *A* is $\frac{1}{2}$.

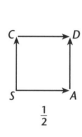

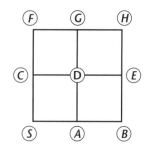

3. What is the probability that Sam goes to *C*? _____

4. What is the probability that Sam goes to *D* from *C*? _____

5. What is the probability that Sam goes to *D* from *A*? _____

6. Find the probability that Sam goes to *D*. _____

7. Find the probability that Sam goes to *F*. _____

Cooperative-Learning Activity
5.1 Fantastic Functions

Group members: 2–4

Materials: pencil

Responsibilities: As a group you will interact with function machines.

Preparation: Below are function machines. When a number goes into a machine, it comes out a changed number—it has been functionized.

Example:

x	y
1	3
−1	−3
4	12
0	0

$$1 \atop \downarrow \rightarrow 3 \qquad -1 \atop \downarrow \rightarrow -3 \qquad 4 \atop \downarrow \rightarrow 12 \qquad 0 \atop \downarrow \rightarrow 0$$

The rule is that for every x value put into the function machine, 3 times that value comes out: $y = 3x$.

Procedures: In pairs, identify the rules for the following functions:

1. $0 \atop \downarrow \rightarrow -3 \qquad 1 \atop \downarrow \rightarrow -2 \qquad 2 \atop \downarrow \rightarrow -1 \qquad 3 \atop \downarrow \rightarrow 0$

   Rule: _____

2. $-1 \atop \downarrow \rightarrow 1 \qquad 2 \atop \downarrow \rightarrow 2 \qquad -4 \atop \downarrow \rightarrow 4 \qquad 8 \atop \downarrow \rightarrow 8$

   Rule: _____

3. $1 \atop \downarrow \rightarrow 1 \qquad 3 \atop \downarrow \rightarrow 9 \qquad -4 \atop \downarrow \rightarrow 16 \qquad -2 \atop \downarrow \rightarrow 4$

   Rule: _____

In pairs, make up two function machines on your own. Then exchange them with another pair and identify each other's rules.

4.

   Rule: _____

5.

   Rule: _____

Cooperative-Learning Activity
5.2 Transformers

Group members: 4

Materials: pencil, a spinner with the numbers 1 to 5, number cubes, and a coin

Roles: **Spinner** spins the spinner to determine the type of function

Flipper flips the coin to determine if the function is positive or negative

Roller rolls the dice to determine the numerical values of the function

Assembler generates the function with the noted specifications

Preparation: As a group, you will construct and graph different types of transformable functions.

Procedures:
1. The Spinner spins the spinner to tell what kind of function the group is to use.

    1: Linear
    2: Exponential
    3: Quadratic
    4: Reciprocal
    5: Absolute value

2. Then the Flipper flips the coin to determine whether the function is positive or negative.

    Heads: Positive
    Tails: Negative

3. The Roller rolls number cubes to obtain the numbers that can be used in the function.

4. The Assembler generates a function with the above specifications, and everyone in the group must write it down and graph it.

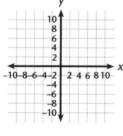
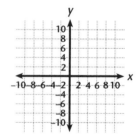

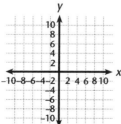
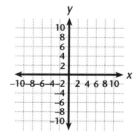

5. Each group member will then state whether the function has been stretched, reflected, or shifted. Rotate the roles so that each group member has a chance to generate a function at least once.

Cooperative-Learning Activity
5.3 Stretch that Function

Group members: 2–4

Materials: pencil and a number cube

Roles: **Roller** rolls the number cube to determine how much to stretch the function

Writer writes down the stretched function

Grapher graphs the stretched function

Preparation: As a group, you will perform stretches on parent functions, and graph the results.

Procedures:
1. Use the following functions in the order given:
$$y = x^2$$
$$y = |x|$$
$$y = \frac{1}{x}$$

2. The Roller rolls the number cube to find out how much to stretch the function.

3. The Writer writes the new function, and the Grapher graphs the function on the coordinate axis given. Rotate roles until each group member has performed in each role.

New function: _____

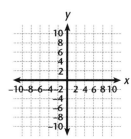

New function: _____

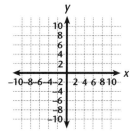

New function: _____

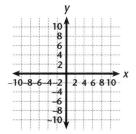

Cooperative-Learning Activity
5.4 Spoons With Reflections

Group members: 4–5

Materials: pencil, 25 3 × 5 cards, and spoons

Roles: **Shuffler** shuffles the cards and places them facedown

Dealer deals the cards facedown and begins passing the cards

Players play spoons

Preparation: Each player will attempt to be the first to match a function with its reflection and attempt to secretly get a spoon.

Procedures:

1. As a group, divide up the following expressions and put each one on a 3 × 5 card.

-2^x	2^x	$x^2 + 3$	$x^2 - 3$	$3x^2$	$3x^2 + 4$	$\left(\dfrac{1}{2}\right)^x$	$-\dfrac{1}{2^x}$
$-\dfrac{1}{4x^2}$	$\dfrac{1}{4x^2}$	$\dfrac{3}{x}$	$-\dfrac{3}{x}$	$\dfrac{8}{x}$	$\dfrac{8 + x}{x}$	$-|x|$	$|x|$
$6x^2$	$-6x^2$	$3x^2 + \dfrac{1}{2}$	$3x^2 - \dfrac{1}{2}$	-8^x	8^x	$\dfrac{1}{x}$	$-\dfrac{1}{x}$

2. Form a circle and set the spoons in the middle of the circle. There should be one less spoon than there are people.

3. After the index cards are labeled, the Shuffler shuffles the cards and places them facedown.

4. The Dealer should deal each person two cards facedown. Each player may look at his or her cards. The object is to form a pair of cards. A pair is a function and its reflection.

5. Decide which card to keep, and pass the card that you want to throw away facedown to the person on the right.

6. The Dealer draws from the pile to introduce new cards, and he or she passes his or her cards to the right. The last person puts his or her cards into a throw away pile, not to be touched again until the next game.

7. The first person to get a pair must try to secretly take a spoon from the center of the circle. If another player notices that a spoon is missing, he or she also takes a spoon. The person who is left without a spoon is the new Dealer.

Cooperative-Learning Activity
5.5 Topsy-Turvy Translations

Group members: 2–4

Materials: pencil and 3 × 5 cards

Responsibilities: Determine translated points from translations.

Preparation: As a group, put each of the following points on a 3 × 5 card and make a pile: (3, 4), (1, 5), (0, 2), (1, 1)

Then put each of the following translations on a 3 × 5 card and make a pile:

Vertical translation by 10	Horizontal translation by 4
Vertical translation by −7	Horizontal translation by −9
Vertical translation by −3	Horizontal translation by $-\frac{1}{2}$
Vertical translation by 25	Horizontal translation by 11
Vertical translation by $\frac{1}{2}$	Horizontal translation by −13

Procedures:
1. The two piles must be kept separate. A player shuffles each pile of cards facedown and puts them in the center of the group.

2. The first player picks a card from the points pile and displays the card faceup next to the pile. Each player in the group will use this point.

3. Then each player picks a translation card. Record the point, the translation, and the translated point in the table below for each player.

Point	Translation	Translated point

4. A new player then picks a point card. When you run out of translation cards, recycle the old ones by shuffling them and starting over.

NAME _____ CLASS _____ DATE _____

Cooperative-Learning Activity
5.6 Building a Transformation

Group members: 2–4

Materials: pencil, 3 × 5 cards, and a number cube

Roles: **Functioner** writes each of the functions on a 3 × 5 card

Transformer writes each of the transformations on a 3 × 5 card

Roller rolls a number cube to determine specific translations and stretches

Dealer shuffles the cards and places them facedown

Preparation: As a group, you will write and graph transformed parent functions.

Procedures: 1. The Functioner will write each of the following functions on a 3 × 5 card:

Exponential	Quadratic
Reciprocal	Absolute value

The Transformer will write each of the following transformations on a 3 × 5 card:

Reflection	Horizontal translation
Vertical translation	Vertical stretch

2. The Dealer shuffles the two sets of cards and places them facedown. Keep each pile separate.

3. The Dealer picks one card from the function pile and two cards from the transformation pile, and the Roller rolls the number cube.

4. Each group member writes a function that satisfies the given transformations.

5. Each player passes his or her function to the right so that another group member can check to see if it is correct by graphing it in the space below.

Function: _____

Function: _____

Function: _____

Function: _____

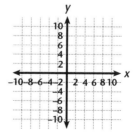

6. Replay the game and rotate roles until each group member has sketched four graphs.

Cooperative-Learning Activity
6.1 Exploring Exponents

Group members: 2

Materials: graph paper and a pencil

Responsibilities: Use a chessboard to explore number patterns.

Preparation: Cut out an 8-inch-by-8-inch grid to model a chessboard.

Procedures: King Kaid of India was looking for new challenges. After many years of work, a loyal subject created the game of chess. As his reward, he received 1 grain of corn for the first square on the chessboard, 2 grains for the second square, 4 for the third square, and so on for the entire 64 squares on the chessboard. The king soon discovered that this reward would be worth more than his whole kingdom.

As a group, find out how many grains of corn would be on the entire chessboard.

1. Before attempting to solve the problem, make an estimate of the total

   number of grains on the chessboard. _____

2. Express each number of grains per square as a power of 2. How many grains

   of corn are on the 4th square? _____

   the 16th square? the 64th square? _____

   As a group, calculate the sum. You can do this by calculating each power of 2 and then adding the results. However, this method is long and time consuming. A more efficient way is to calculate the sum and look for a pattern. Examine the partial sums:

   $$1 = 1$$
   $$1 + 2 = 3$$
   $$1 + 2 + 4 = 7$$
   $$1 + 2 + 4 + 8 = 15$$
   $$1 + 2 + 4 + 8 + 16 = 31$$

3. Describe what you notice about the sum in each row and the last addend

   in the next row. _____

4. Use this pattern to find the sum of all the grains on the chessboard. Write an

   estimated name for the sum. _____

5. Compare your group's estimate with the actual result.

   Did your group overestimate or underestimate? _____

Cooperative-Learning Activity
6.2 Chessboard Calculator

Group members: 2

Materials: markers, grid paper, or a chessboard

Responsibilities: Explore how a chessboard can be used to model the product of two numbers.

Preparation: Cut out an 8-inch-by-8-inch grid to model the chessboard. Write the numbers along the horizontal and vertical columns as shown in the diagram.

Procedures: John Napier, a mathematician, developed a chessboard calculator based on the binary system (base 2).

1. Use the model below to multiply 13 × 11. Express the number 13 with markers along the bottom row at 8, 4, and 1 because 8 + 4 + 1 = 13, and the number 11 with markers along the right vertical column at 8, 2, and 1 because 8 + 2 + 1 = 11.

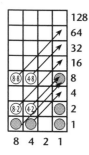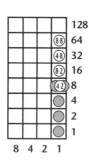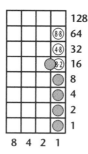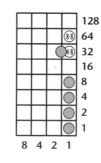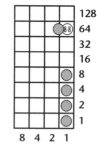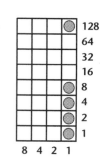

2. Place a new marker at the intersecting square where a row with a marker meets a column with a marker.

3. To multiply, slide the markers from the bottom row and the markers at the intersection diagonally to the vertical row.

4. Examine the vertical column. Remove any two markers sharing the same square and place one marker in the square above them. Repeat this replacement if needed.

5. The vertical column that results will represent the product of 13 × 11.

Use Napier's chessboard calculator to find each product.

6. 8 × 32

7. 12 × 12

8. 14 × 6

9. 64 × 16

Cooperative-Learning Activity
6.3 Paper Folding and Exponents

Group members: 4

Materials: 12 sheets of newsprint, paper, and a pencil

Roles: **Leader** asks the teacher for assistance, if needed

Folder folds newsprint following the procedures

Recorder records group answers in the given table

Counter counts sections in the newsprint created by the Folder

Preparation: Use paper folding to model exponential decrease.

Procedures:
1. The Folder takes a large sheet of newsprint, folds it in half horizontally, and then unfolds it. The Counter counts the sections in the newsprint created by the folds. The Folder refolds the newsprint. As a group, describe the change in area of the folded sheet and compare it with the unfolded sheet. The Recorder writes the changes on a piece of paper.

2. The Folder now folds the newsprint in half again and then unfolds it. The Counter counts the number of sections created. Refold the newsprint. As a group, describe the change in area of the newsprint as a result of the additional fold. The Recorder writes the changes on a piece of paper.

3. The Folder continues the process of folding and unfolding the newsprint. As a group, determine the change in area caused by each new fold. The Recorder records the group's results in the table.

Number of folds	1	2	3	4	5					
Area of folded paper										

4. As a group, examine the data in the table. Use the pattern you observe to extend the table beyond the values that are shown, and write a formula describing this pattern. As a group, describe the change in area as the number of folds increases. The Recorder writes the changes on a piece of paper.

5. As a group, describe what you think will happen if you fold the newsprint into thirds. Then fold the newsprint into thirds. Guess the change in area. Unfold the newsprint, check your guess, and record your results in a table similar to the one above. Refold the newsprint. Fold the newsprint into thirds a second time and guess the change in area. Unfold the paper, check your guess, and record the results. Continue the process of folding and unfolding the paper, and record the results in your table.

6. As a group, describe the similarities and differences between folding newsprint in half and in thirds.

Cooperative-Learning Activity
6.4 Measurement Makers

Group members: 4

Materials: 4 yardsticks, tape, paper, and a pencil

Roles: **Leader** counts the students standing in the square and asks the teacher for assistance, if needed

Recorder records the group's results of estimation versus actualization

Measurer measures a square and the dimensions of a room

Marker marks areas that are measured

Preparation: In the real world, there are situations in which it may not be practical to solve a problem by counting or using a formula. A more practical method may be estimation.

Procedures: 1. Imagine that your classroom is totally empty. There are no people, desks, or furniture. As a group, estimate the number of students that it would take to completely cover the floor of your classroom. The Recorder records the estimate. Discuss the possible methods your group could use to solve this problem.

2. Do you think that different methods would yield the same answer? Explain.

_____

Use the following method to solve this problem:

3. In an open area of the classroom, the Measurer measures a square with a yardstick. The square should have a dimension of 1 yard on each side. The Marker marks the sides of the square with tape. Select group members to line up inside the square, shoulder-to-shoulder, along the length and front-to-back along the width until the square is completely filled. The Leader counts the number of students standing inside the square, and the Recorder records the number.

_____

4. Briefly examine the room. Estimate the number of squares (1 yard × 1 yard) in the room. Then the Measurer measures the dimensions of the room and determines the area. How accurate was your group's estimate? The Recorder records the measurement.

_____

5. Multiply the number of students per square yard by the number of squares in the room. How many students can stand inside your classroom? The Recorder records the group's result.

_____

6. Compare the calculated result to the group's original estimate. Did your group overestimate or underestimate? Explain.

_____

Cooperative-Learning Activity
6.5 Exponential Functions

Group members: 2

Materials: counters, grid paper, and a pencil

Responsibilities: Examine patterns and strategies.

Preparation: Enlarge the game board below.

Procedures:
1. Place three counters on the numbers 1, 2, and 3 in Sector A. Move the counters in the fewest possible moves so that they all end up in Sector B. In order to move a counter, the following conditions must be met:

    a. Counters are moved one at a time, each within its own ring, to either Sector B or Sector C.

    b. A counter may not be moved to an adjacent one in which there is a counter closer to the center than the one being moved.

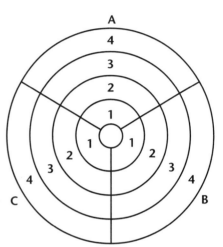

2. Work in pairs, with one partner recording the moves. Continue playing until you can transfer the markers in 7 moves. As you analyze your moves for each game, pay attention to the counter on the smallest ring.

3. Now play the game with 4 markers. The minimum number of moves required for 4 markers is 15. As you make your moves, record them. Then analyze each move that you made to reach 15.

4. Playing the game with 5 markers can be difficult, so look for a pattern in the games involving 1, 2, 3, and 4 markers. Complete the table and predict the number of moves required to move 5 counters.

Number of markers	1	2	3	4	5
Number of moves					

5. Graph the data from your table. Compare your graph to the graph of $y = -2^x$. How are the graphs alike? How are the graphs different?

6. Use your graph to determine the minimum number of moves needed to transfer 5 counters.

7. Determine the minimum number of moves needed to transfer n coins.

Cooperative-Learning Activity
6.6 Water Mixtures and Exponential Functions

Group members: 4

Materials: tap water, ice water, hot water, effervescent tablets (dental or aspirin), Celsius thermometer, 5 clear plastic cups, clock with a second hand, measuring cup, and a pencil

Responsibilities: Conduct an experiment to model exponential decay.

Procedures: 1. Pour 5 ounces of ice water in one cup, 5 ounces of tap water in another cup, and 5 ounces of hot water in a third cup.

2. In a fourth cup, mix 2.5 ounces of tap water with 2.5 ounces of ice water. In the fifth cup, mix 2.5 ounces of tap water with 2.5 ounces of hot water.

3. Measure and record the temperature in each cup.

4. One member acts as a timekeeper and instructs the other members to drop an effervescent tablet into each cup at the same time.

5. Record the time in seconds when the tablet stops fizzing.

Temperature in °C (x)	
Time in seconds (y)	

6. Graph your data on the given coordinate axis.

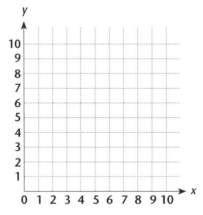

7. Use the graph to extend the pattern. Predict the amount of time necessary for the reaction to be completed at a temperature that is halfway between

the hot-water temperature and the tap-water temperature. _____

8. Compare your graph with the graph of $y = 2^x$. How are the graphs alike? How are they different?

_____

9. As a group, determine the relationship between temperature and reaction time.

_____

Cooperative-Learning Activity
7.1 Magazine Subscriptions

Group members: 2

Materials: graphics calculator and a pencil

Roles: **Recorder** creates a chart and records the group's answers

Checker uses a graphics calculator to verify answers

Preparation: Each year a magazine publisher sells annual magazine subscriptions to students through the Student Council. The subscription price is $27.50. The publisher expects to sell 100 subscriptions at that price. However, they want to sell more. To increase sales, they decide to make a special offer in which they will reduce the subscription price by $0.20 for each subscription over 100.

Procedures: 1. To determine how this progressive price reduction can affect the publisher's revenue, the Recorder creates and completes a chart similar to the one below, extending the subscriptions to 121. As the Recorder fills in the chart, the Checker will calculate the total revenue. Reverse roles when the number of subscriptions is 112.

Number of subscriptions	Subscription price ($)	Total revenue ($)
100	27.50	2750.00
101	27.50 − (0.20 × 1)	2757.30
102	27.50 − (0.20 × 2)	2764.20
103		
104		
105		

2. From the publisher's point-of-view, does reducing the price in this way make sense? Explain.

_____

3. From the chart, what number of subscriptions sold will produce the maximum revenue for the publisher? At what number of subscriptions should the publisher discontinue the price reduction?

_____

4. Let x represent the number of subscriptions in excess of 100, and let $f(x)$ represent the total revenue. Write a polynomial function that models the data.

_____

Cooperative-Learning Activity
7.2 Algebra Tiles and Polynomials

Group members: 2–3

Materials: grid paper, algebra tiles, and a pencil

Responsibilities: Use algebra tiles to represent a pattern.

Preparation: Your group will use algebra tiles to represent a pattern. As a group, regroup the tiles and use them as a model for polynomial expressions to represent the number of tiles in the nth figure. Examine the tiling pattern. How many tiles will be needed to build the 20th figure in this pattern?

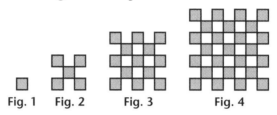

Fig. 1   Fig. 2   Fig. 3   Fig. 4

One way to solve this problem is to regroup the tiles in each figure and look for a pattern. Examine the diagram. Notice how the tiles were regrouped.

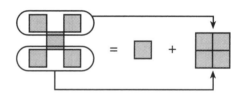

Procedures: 1. Regroup the tiles in the third figure in a pattern like that above. How were the tiles moved to form squares? Count the number of tiles in each square. How many tiles does the third figure have?

_____

2. Continue the process for the fourth figure. How many tiles does the fourth

   figure have?   _____

3. Find a polynomial expression to represent the pattern as shown by this

   regrouping.   _____

4. Use the polynomial expression to predict the number of tiles in the

   20th figure.   _____

5. Regroup the tiles in another way. Write the polynomial expression modeled by this regrouping. Compare this polynomial expression with the expression from step 4. Are the expressions equivalent? Explain.

_____

_____

Cooperative-Learning Activity
7.3 Exploring Multiplication Models

Group members: 2–3

Materials: grid paper and a pencil

Responsibilities: Find polynomial expressions that model a problem.

Preparation: Your group will solve a problem by recognizing a similar pattern in a less complex situation and then finding a polynomial expression that models both problems.

Examine the 4-row-by-3-column rectangular figure. To find the total number of squares in the figure, start with a simpler figure, organize the data, and look for patterns.

Procedures: 1. This figure is a 2-row-by-3-column rectangle. How many of each type of square can you find? How many are in a 1-row-by-1-column rectangle? in a 2-row-by-2-column rectangle? how many unit squares in a 2×2 rectangle?

2. Examine this 3-row-by-3-column rectangle. Find the number of each type of square. Organize the data and record it in the table below.

Number of Squares

Size	1 row by 1 column	2 rows by 2 columns	3 rows by 3 columns	Total
1 row by 3 columns				
2 rows by 3 columns				
3 rows by 3 columns				
4 rows by 3 columns				

3. Record the data for each type of square in a 1-row-by-3-column rectangle.

4. Notice the data. What appears to be the relationship between the size of the rectangle and the number of squares in 1 row by 1 column? Write a polynomial expression to represent this relationship. _____

5. Write a polynomial expression to represent the relationship between the size of the rectangle and the number of squares in 2 rows by 2 columns.

_____

Cooperative-Learning Activity
7.4 Geoboards

Group members: 2–3

Materials: geoboards of different sizes (1 × 1, 2 × 2, 3 × 3, 4 × 4, and 5 × 5) or square dot paper, rubber bands, ruler, and a pencil

Responsibilities: Explore relationships by using geoboards.

Preparation: Your group will use a geoboard to explore the relationship between the size of the geoboard and the number of different line segments lengths that can be created on the geoboard.

Examine the 3 × 3 geoboard shown. Five different line segment lengths have been made.

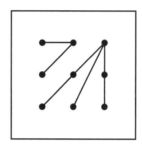

Procedures: **1.** Use rubber bands to make different line segments on geoboards of different sizes. Count the number of line segments for each size geoboard and record this number in the table.

Geoboard size	1 × 1	2 × 2	3 × 3	4 × 4	5 × 5	. . .	$n \times n$
Number of different line segment lengths							

2. Look at the data. What appears to be the relationship between the size of the geoboard and the number of different line segment lengths that can be

made on the geoboard?_____

3. Use the pattern you notice to predict the number of different line segment

lengths that can be made on a 6 × 6 geoboard. _____

4. Let n represent the size of an $n \times n$ geoboard. Represent the number of different line segment lengths as the product of two binomials in terms of n.

_____

5. Use the Distributive Property to write your formula as a trinomial. Then use the trinomial to determine the number of different line segment lengths

possible on a 10 × 10 geoboard. _____

Cooperative-Learning Activity
7.5 Common Factors

Group members: 2

Materials: 2 different colored counters (4 of each color) and a pencil

Responsibilities: Explore common factors.

Preparation: In this activity, you will play a game and then analyze it to determine the minimum number of moves possible.

Procedures: 1. Enlarge the game board shown.

1	2	*empty*	4	5

2. Place a black counter on numbers 1 and 2. Place a white counter on numbers 4 and 5. The object is to reverse the order of the counters in the fewest possible moves, following these conditions:

    a. Counters can be moved forward, one at a time, to an empty space.
    b. Counters can be moved forward, one at a time, jumping over one counter of another color to an empty space.

3. Play in pairs, with one partner recording the moves. Continue playing until you can exchange the colors in eight moves.

4. Now play the game with three black and three white markers. Enlarge the game board shown.

1	2	3	*empty*	5	6	7

Place a black marker on numbers 1, 2, and 3.

Place a white marker on numbers 5, 6, and 7. The minimum number of moves required to exchange the three markers of each color is 15. As you make your moves, record them. Then analyze the outcome to reach 15.

5. Suppose that you play the game with four markers of each color. What is the minimum number of moves needed to exchange the colors? Look for a pattern in the games involving one, two, and three markers. Complete the table. Then predict the minimum number of moves needed to exchange four markers of each color.

Number of markers of each color, n	1	2	3	4
Minimum number of moves				

6. Examine the data: $1(1 + 2) = 3$
   $2(2 + 2) = 8$
   $3(3 + 2) = 15$

Use this pattern to write the polynomial that describes the minimum

number of moves. _____

NAME _____ CLASS _____ DATE _____

Cooperative-Learning Activity
7.6 Tessellation's

Group members: 3

Materials: tessellation tracer or templates of regular hexagons and a pencil

Roles: **Counter** counts vertices in a tessallation

Recorder records the group's answers in the given table

Drawer draws tessellations described in procedures

Preparation: A tessellation is a pattern in which congruent copies of a figure completely fill the plane without overlapping. In this activity, your group will design hexagonal tessellations and then explore the relationship between the size of the design and the number of vertices in the design.

Procedures:
1. As a group, examine this 2×2 hexagonal tessellation. The Counter counts the number of vertices and the Recorder records the answer in the table below.

2 x 2

Design	1×1	2×2	3×3	4×4	. . .	$n \times n$
Number of vertices						

2. The Drawer draws a 3×3 hexagonal tessellation. The Counter counts the number of vertices, and the Recorder records this number in the table above.

3. As a group, work backward by using the pattern, and predict the number of vertices in a 1×1 hexagonal tessellation. The Recorder records the result.

4. Do you see a pattern? Examine the factors of the number of vertices and look for a pattern. Then, as a group, fill in the missing numbers.

Design	Number of vertices
1×1	$6 = 2 \times 1 \times 3$
2×2	_____ $= 2 \times 2 \times 4$
3×3	_____ $= 2 \times 3 \times$ _____
4×4	_____ $= 2 \times 4 \times$ _____

5. Write the polynomial that models this pattern. Then use this polynomial to predict the number of vertices in a 9×9 hexagonal tessellation.

_____

Cooperative-Learning Activity
7.7 Models and Trinomials

Group members: 3–4

Materials: 1-tiles, x-tiles, x^2-tiles, and a pencil

Responsibilities: Use models to factor trinomials.

Preparation: In this activity, members of the group will find the greatest number of ways to arrange algebra tiles to form a rectangle. The area of the rectangle represents a quadratic trinomial, and the length and width represents the factored form of the quadratic trinomial.

Procedures:

1. Each team member is given 6 x^2-tiles, 15 positive x-tiles, 15 negative x-tiles, 12 positive 1-tiles, and 12 negative 1-tiles.

2. Each member must create a trinomial according to these conditions:

   a. The x-tiles must be placed with the long side along the edge of an x^2-tile or the long side of another x-tile.
   b. The 1-tiles must be placed at the end of an x-tile or next to another 1-tile.
   c. The model of the trinomial must form a rectangle.
   d. The length and width of the rectangle must represent the factored form of the trinomial.

3. Examine the tiles that model $2x^2 + 5x + 2 = (2x + 1)(x + 2)$. Describe what you notice.

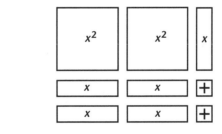

_____

_____

_____

4. Build trinomials. Record the tile patterns in the space provided below. Then check the trinomial by multiplying the factors.

Cooperative-Learning Activity
8.1 Patterns From Folding a Circle

Group members: 2–3

Materials: compass, ruler, scissors, paper, and a pencil

Responsibilities: Collect and organize data by folding a circle.

Preparation: In this activity, you will construct a circle, fold it to generate data, and examine the data by using differences.

Procedures:
1. Each group member will construct a circle with a 3-inch radius on paper and cut it out.

2. Each member will then attempt to maximize the number of sections created on their circle by a series of folds. Try to predict the number of sections you can create with each fold of the circle.

3. After making each fold, unfold the circle and record the total number of sections created in the table below. Members should compare their results to determine the maximum number of sections.

Number of folds, x	0	1	2	3	4	5	6
Number of sections, $f(x)$							

4. Remember to open the circle before each additional fold, and always fold for the maximum number of sections possible. Complete the table up to 4 folds.

5. As a group, examine the data. Calculate the first differences for $f(x)$ and describe what you notice. _____

6. Calculate the second differences for $f(x)$. _____

7. Explain how you can tell from second differences that the values of this function represent a quadratic. _____

_____

8. As a group, work backward by using the pattern and predict $f(5)$ and $f(6)$.

_____

_____

Cooperative-Learning Activity
8.2 Peas and Parabolas

Group members: 4–6

Materials: several circular jar lids of different sizes, ruler, bag of dried peas, container, grid paper, and a pencil

Responsibilities: Collect data individually, and draw a graph of the group data.

Preparation: Each group will determine the relationship between the radius of a circular lid and the number of peas that fill the bottom of the lid.

Procedures:

1. Each member of the group chooses a different size lid. Measure the diameter of each lid. Then determine the radius and record the results in the table below.

Radius of lid (in.)	Number of dried peas

2. Empty the bag of dried peas into the container.

3. Fill the bottom of the lid with dried peas. Make sure that no pea lies on top of another. Count the number of peas and record the number in the table above.

4. Form ordered pairs from the data in the chart and display the information as a scatter plot with the radius of the lid on the horizontal axis and the number of dried peas on the vertical axis on grid paper.

5. Connect the points with a smooth curve. As a group, compare your graph with the graph of $y = x^2$. How are the graphs alike? How are they different?

   _____

6. Use your group's graph to predict the number of dried peas that would fill a

   lid with a radius of 12 inches. _____

7. Describe the relationship between the radius of the lid and the number of

   peas that fill the bottom. _____

Cooperative-Learning Activity
8.3 Using Tiles to Complete the Square

Group members: 4

Materials: algebra tiles, paper and a pencil

Roles: **Leader**   asks teacher for assistance, if needed

Arranger   arranges the tiles to form each figure

Counter   counts the number of tiles in each figure

Recorder   records the group's answers

Preparation: Your group will use algebra tiles to predict the number of tiles in the *n*th figure by recognizing a pattern and creating a model of a quadratic expression.

Procedures:

1. As a group, examine the pattern. The Arranger arranges the tiles to form the first figure. The Counter counts the number of tiles. Repeat the process for the second figure. The Recorder records the answers to the following questions: How many tiles are in the first figure? How many tiles are in the second figure?

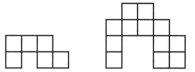

2. Continue the process for the third figure. How many tiles are in the third figure? Does the method of counting easily lead to a pattern for the number of tiles in the *n*th figure?

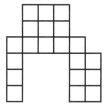

3. The Arranger arranges the tiles to form the third figure. Now move the tiles in a pattern like one shown below. As a group describe the result. The Recorder records the description.

4. As a group, work backward by using the pattern and predict the number of tiles needed for the first figure.

5. As a group, find a general formula for the number of tiles in the *n*th figure.

Cooperative-Learning Activity
8.4 Modeling the Stream From a Water Fountain

Group members: 4

Materials: ruler, wire clothes hanger, wire cutter, paper towels, paper, and a pencil

Roles: **Recorder** records the group's results

Designer creates a model of a stream from a water fountain

Measurer measures the waterspout

Verifier uses a graphics calculator to create a 2-variable data table

Preparation: Your group will find a mathematical function that represents a stream of water from a water fountain.

Procedures:
1. Turn on the water fountain so that the water flows at a constant rate. Keep the water flowing so that an arch forms. What kind of curve does the

   stream of water model? The Recorder records the group's answer. _____

2. The Designer creates a model of the stream of water by using a wire clothes hanger.

3. The Measurer measures the water spout of the water fountain and the pan

   into which the water falls.   _____

4. The Recorder makes coordinate axes and then positions the wire hanger so that it models the stream coming from the fountain.

5. As a group, estimate the x-intercepts of the wire hanger. Round to the

   nearest tenth of an inch, if necessary.   _____

6. Determine the maximum y-value of your model.   _____

7. Using a graphics calculator, the Verifier enters the x-intercepts and maximum y-value, as ordered pairs, into a two-variable data table. Then the Verifier fits the data to the second-order polynomial $y = ax^2 + bx + c$ by selecting QuadReg.

8. The Recorder writes the quadratic equation that models the stream of water

   from the fountain.   _____

9. As a group, describe how to test your equation to determine how well it approximates the stream of water flowing from the fountain.

   _____

   _____

   _____

Cooperative-Learning Activity
8.5 Quadratic Patterns

Group members: 2–4

Materials: rectangular tiles and a pencil

Responsibilities: Solve a problem by using similar patterns.

Preparation: Your group will solve a problem by recognizing a similar pattern in a less complex situation and then finding a quadratic function that models both problems.

Procedures: **1.** Examine the figure. Suppose that you were asked to find the total number of rectangles formed within the figure. Describe the method you would use to solve the problem.

_____

_____

To find a formula that will solve this problem and any similar problem, start with a simpler problem.

2. Use rectangular tiles. Assign each tile a letter: A, B, C, etc. Start with one tile, A. Next add a second tile, B. How many rectangles are in each figure? Hint: List the number of rectangles you can find by letter.

3. Continue this process by adding a third tile, C. How

many rectangles are in this figure? _____

4. Record the data in the table below.

Number of small rectangles, x	1	2	3	4	5
Total number of rectangles, $f(x)$					

5. Do you see a pattern? Notice the relationship between the total number of

rectangles and the number of small rectangles. _____

6. As a group, explain how the values of this function represent a quadratic

function, work backward by using the pattern, and predict $f(8)$ and $f(10)$.

_____

_____

Cooperative-Learning Activity
8.6 Disk Toss

Group members: 5–7

Materials: circular disks of various sizes, centimeter ruler, tape, 9-inch by 9-inch square tiles, grid paper, and a pencil

Responsibilities: Use an experiment to find the quadratic function of best fit.

Preparation: Each group will compare the results of tossing different-size circular disks from a predetermined distance onto a square consisting of 9-inch by 9-inch tiles. The objective is to toss the disk so that it lands completely inside a 9-inch by 9-inch tile, not in contact with any of the edges of any tile. A successful toss gains one point.

Procedures:

1. Lay a square on the floor consisting of 9-inch by 9-inch tiles. Then make a line with tape a measured distance away from the square.

2. Choose a different size circular disk for each member. Measure the diameter of the disks and record this data for each member of your group.

3. Each player will stand behind the line and toss a disk onto the square. Each player will continue to use the same disk for each round and will make a minimum of 50 tosses. Record the number of successful tosses for each player.

Type of disk	Diameter (mm)	Number of successful tosses	Number of successful tosses / total number of tosses

4. For each type of disk, calculate the ratio of the number of successful tosses to the total number of tosses, and record the answer in the table.

5. Form ordered pairs from the data in the chart. The x-coordinate is the diameter of the disk, and the y-coordinate is the ratio of the number of successful tosses to the total number of tosses. As a group, display the information as a scatter plot on grid paper. Then draw a curve so that it connects most of the data points. Does the graph appear to be quadratic? Explain the group's answers on a piece of paper.

6. Use a graphics calculator to find the quadratic function of best fit.

_____

Cooperative-Learning Activity
9.1 Modeling Windchill With Square-Root Functions

Group members: 2–3

Materials: graphics calculator and a pencil

Responsibilities: Perform calculations with piecewise square-root functions to complete a table that models windchill.

Preparation: Wind can make the temperature feel much colder than it actually is because wind intensifies the loss of heat from the skin. The windchill is what the temperature would have to be with no wind in order to give the same chilling effect.

Procedures: Examine the model for finding the windchill temperature, T_w, where t is the actual temperature in degrees Fahrenheit, given by a thermometer and v is the wind speed in miles per hour.

$$T_w = \begin{cases} t & 0 \leq v \leq 4 \\ 0.0817(5.81 + 3.71\sqrt{v} - 0.25v)(t - 91.4) + 91.4 & 4 < v \leq 45 \\ 1.60t - 55 & v > 45 \end{cases}$$

1. This mathematical model represents a piecewise function. As a group, explain why $T_w = t$ when $0 < v < 4$.

   _____

2. As a group, identify the piece of information which represents that wind speeds greater than 45 miles per hour do not significantly reduce body heat.

   _____

3. Take turns using a graphics calculator to find the windchill temperatures. Round degrees to the nearest tenth. Then record the results in the table.

Windchill Temperature

Wind speed	Temperatures in degrees Fahrenheit								
	−20°	−10°	−5°	0°	5°	10°	20°	30°	40°
10 mph									
15 mph									
20 mph									
25 mph									
30 mph									
40 mph									

Cooperative-Learning Activity
9.2 Approximating the Reciprocals of Square Roots

Group members: 2–3

Materials: calculator, paper, pencil, and a ruler

Roles: **Drawer** draws the ladders on paper for the group to work with

Recorder writes the missing numbers in the ladder

Calculator calculates the ratios of the numbers connected at side end of the ladder

Preparation: Your group will explore the ladder arithmetic used by the Greeks to approximate the value of the reciprocal of square roots.

Procedures: Follow these steps to draw the ladder, and use the ratios that the ancient Greeks used to approximate the value of $\frac{1}{\sqrt{2}}$:

1. The Drawer copies the ladder on a separate piece of paper.

Number ladder

2. The Recorder writes the number 1 at the top of the ladder in the first column. Then the Recorder writes the number 1 at the top of the ladder in the second column.

3. Now, as a group, determine the other numbers to be placed on the ladder. Examine how the numbers are generated in each column of the ladder. Using this pattern, the Recorder writes in the missing numbers.

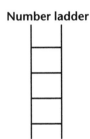

Number ladder

1	1	$\frac{1}{1}$ = 1
2	3	$\frac{2}{3}$ = 0.6666
5	7	$\frac{5}{7}$ =
12		
	41	

4. As a group, find the ratio of the numbers connected at each side of the ladder. As a group, complete the chart on a separate piece of paper. The Calculator will use a calculator to verify the group's answers.

Number	Ratio
$\frac{1}{1}$	1
$\frac{2}{3}$	0.6666. . .

5. The Calculator will use the graphics calculator to approximate $\frac{1}{\sqrt{2}}$. As a group compare this result with the ratios determined by the ladder. As a group, describe what you notice.

6. Using a ladder similar to the one in Procedure 1, take turns calculating the ratios of the numbers on the same steps of the ladder. The numbers 2 and 2 are in the top of the ladder. Then, as a group, determine the irrational numbers that the ratios approximate.

Cooperative-Learning Activity
9.3 The Golden Pentagram Star

Group members: 4

Materials: graphics calculator, pencil, ruler, stiff cardboard, needle and thread, tape or glue, and scissors

Roles: **Solver** solves equations by using the quadratic formula

Cutter cuts cardboard to model a golden rectangle

Taper tapes the frame to strengthen it

Sewer sews corners to construct the golden pentagram star

Preparation: Your group will construct a golden frame. Then you will use the frame to construct a pentagram star.

Procedures: The divine proportion, or golden ratio, was derived by fifteenth-century mathematician, Luca Pacioli. Segments are in the divine proportion if the following is true:
$$\frac{1 - x}{x} = \frac{x}{1}.$$

1. The Solver will solve this equation for a positive value of x using the quadratic formula. Calculate the approximate value of the golden ratio

   with a calculator. _____

2. Now you will construct a golden frame. The Cutter will cut a sheet of stiff cardboard into a rectangle measuring 25 centimeters by 15.5 centimeters. As a group, examine the rectangle and explain why this rectangle is called a golden rectangle.

3. The Cutter will cut two more identical rectangles. The Cutter then cuts slits along the center of the golden rectangle's as shown. The slits should be long enough to slip another rectangle's smaller side through. Extend the slit on the third rectangle to one edge.

   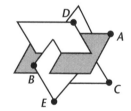 

4. The Taper will form the golden frame as shown in the diagram. Then the Taper will run tape along the edges and where the cards meet to strengthen the frame.

5. Using a needle and thread, the Sewer will sew through five corners (A, B, C, D, and E) to construct a golden pentagram star.

6. You can use the frame to create an octahedron. Repeat Procedures 2–5. Then locate and mark the midpoints of the short sides. Using a needle and thread, connect the midpoints of the short sides of each card.

Cooperative-Learning Activity
9.4 Distance on a Geoboard

Group members: 2–3

Materials: 3 × 3 geoboard, rubber bands, and a pencil

Responsibilities: As a group, you will investigate distances by using a geoboard.

Preparation: You will use a geoboard to investigate the distance between any possible pair of pegs. Examine your 3 × 3 geoboard. Then use tape to label the pegs with letters, as shown in the diagram.

A B C
D E F
G H I

Procedures:

1. As a group, list and count the number of different segments that you can construct with peg *A* as an endpoint. _____

2. As a group, list and count the number of different line segments that you can construct with peg *B* as an endpoint, but do not include peg *A*.

   _____

3. As a group, list and count the number of different line segments that you can construct with peg *C*, but do not include pegs *A* or *B*. _____

4. Continue this process to complete the list of line segments. Then, as a group, determine the total number of different line segments possible.

   _____

5. Take turns to complete a chart like the one below, listing all of the possible line segments. Use the "Pythagorean" Right-Triangle Theorem to find the length of the line segments.

	A	B	C	D	E	F	G	H	I
A									
B									
C									
D									
E									
F									
G									
H									
I									

6. As a group, examine the chart. Describe the patterns you notice.

   _____

7. The distance from *A* to *I* is twice as long as the distance from *A* to *E*. As a group, describe how you can use this distance to show that

   $2\sqrt{2} = \sqrt{8}$. _____

8. Extend the pattern for a 3 × 3 geoboard to a 4 × 4 geoboard. As a group, determine the total number of different line segments that can be

   constructed on this board. _____

Cooperative-Learning Activity
9.5 Number of Triangles and the Distance Formula

Group members: 2

Materials: graph paper, ruler, and a pencil

Roles: **Drawer**   draws points on graph paper

Recorder   lists and counts triangles constructed by drawer

Preparation: Your group will determine the number of different triangles that can be made by choosing 3 of 7 given points.

Procedures: As a group, examine the diagram to find how many different triangles can be made by choosing three of the given points: *A, B, C, D, E, F,* and *G.* To solve this problem, start with a simpler problem.

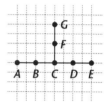

1. The Drawer draws points *A, B, C, D, E, F,* and *G* on graph paper.

2. As a group determine the number of different triangles you can construct using each pair of points as vertices. The Recorder lists the group's results.

    A and *B*   *A* and *C*   *A* and *D*   *A* and *E*   *A* and *F*

    _____   _____   _____   _____   _____

3. As a group, count the number of different triangles that you can construct with point *B.* The Recorder lists the results. Do not count the triangles that include vertex *A.* How many triangles did your group construct?

    _____

4. Continue this process to complete the list of triangles. Then, as a group, determine the total number of triangles.

    _____

5. Complete the chart as a group. Do triangles with the same area have the same perimeter? Explain.

Triangle	Perimeter	Area
△ACG		
△BDG		
△AEF		

6. As a group, decide which description applies to △*FGB*: scalene, isosceles, or

    right. _____

Cooperative-Learning Activity
9.6 Exploring Geometric Properties

Group members: 2–3

Materials: compass, ruler, paper, and a pencil

Roles: **Sketcher** sketches the graph of each given circle

Marker marks and names the intersections

Recorder records the group's results

Preparation: As far back as 1800 B.C. people have tried to solve the problem known as "squaring the circle." The challenge was to construct a square whose area or perimeter was equal to the area or circumference of a given circle. Many people have tried to solve this problem over the centuries. Even though a proof exists that shows the construction cannot be made with Euclidean tools, people still try to construct a "quadrature of the circle." In this activity, you will construct a square nearly equal in area to the area of a circle.

Procedures: **1.** On the same set of coordinate axes, the Sketcher sketches the graph of each circle with the given center and radius.

        a. $(1, 0); r = 1$ **b.** $(0, 1); r = 1$
        c. $(-1, 0); r = 1$ **d.** $(0, -1); r = 1$

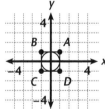

2. As a group, determine where the four circles overlap. Then the Marker marks and names the intersection points A, B, C, and D. Next the Sketcher draws AB, BC, CD, and DA. This is the "quadrature of the circle."

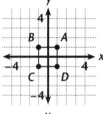

3. The Sketcher sketches the graph of the circle that has point $(0, 0)$ as the given center and a radius of length 1. This is the circle whose area nearly equals the area of square $ABCD$.

4. As a group, describe a way to determine if the area of square $ABCD$ nearly equals the area of circle P. The Recorder records the group's description on a separate piece of paper.

5. Given points $A\left(\dfrac{-1 + \sqrt{7}}{2}, \dfrac{\sqrt{8 - 2\sqrt{7}}}{2}\right)$, $B\left(\dfrac{-1 - \sqrt{7}}{2}, \dfrac{\sqrt{8 - 2\sqrt{7}}}{2}\right)$,
$C\left(\dfrac{-1 - \sqrt{7}}{2}, \dfrac{-\sqrt{8 - 2\sqrt{7}}}{2}\right)$, and $D\left(\dfrac{-1 + \sqrt{7}}{2}, \dfrac{-\sqrt{8 - 2\sqrt{7}}}{2}\right)$, use the distance formula to find the lengths of the four sides of square $ABCD$ as a group. Then find the area of square $ABCD$.

6. As a group, find the area of circle P. Then compare the area of square $ABCD$ with the area of circle P. Is square $ABCD$ approximately the quadrature of the circle? Explain.

Cooperative-Learning Activity
9.7 The Tangent Function

Group members: 3–4

Materials: graphic calculator, paper and a pencil

Responsibilities: As a group, you will determine the proper distance that a ladder should be placed from its base to the building. You will then determine the proper length of the ladder based on its vertical height from the ground level.

Preparation: Placing an extension ladder against a building seems like a simple task. Yet each year people are seriously injured because the ladder is placed either too close or too far away from the building. Leading experts recommend that the base of the ladder must make a 70° angle with the building.

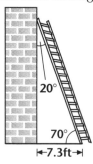

20°

70°

┠←7.3ft→┨

Procedures: 1. As a group, find tan 70°. Use this value and the vertical height of the building to determine the distance from the base of the ladder to the wall. Then use the "Pythagorean" Right-Triangle Theorem to determine the length of the ladder. As a group, take turns completing the chart on a separate piece of paper.

Vertical height (ft)	Base of the ladder to the wall (ft)	Ladder length should be at least (ft)
12		
14		
16		
20		
24		
28		
32		
34		
38		
44		

2. As a group, explain why you think an extension ladder is recommended when the ladder is over 20 feet in length.

_____

Cooperative-Learning Activity
9.8 Triangles and Trigonometry

Group members: 4

Materials: paper clip, protractor, ruler and paper

Roles: **Spinner/Checker**   spins the spinner to determine the known angle and selects the length of the hypotenuse

Sine Calculator   calculates the length of one of the sides of the triangle defined by the Spinner

Cosine Calculator   calculates the length of the other side of the triangle defined by the Spinner

Drawer   draws a triangle to scale by using a protractor and ruler

Preparation: As a group, you will generate different triangles, perform trigonometric calculations to find missing information, and draw a triangle to scale.

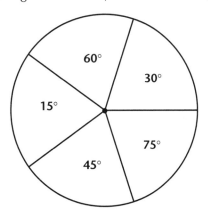

To use the spinner, hold a pencil through one end of a paper clip, and place the pencil point at the center of the spinner. Then spin the paper clip around the pencil like an arrow. Spin again if the paper clip lands in between two degree measures.

Procedures: **1.**   The Spinner uses the spinner to select an angle measure. Then the Spinner chooses a number for the hypotenuse of the triangle.

2.   The Sine and Cosine Calculators find the respective lengths of the missing sides of the triangle defined by the Spinner.

3.   The Drawer makes a scale drawing of the triangle, using the information from the Spinner and Calculators. The Sine and Cosine Calculators check the scale drawing for accuracy.

4.   The Spinner checks the calculations of the Sine and Cosine Calculators by using the "Pythagorean" Right-Triangle Theorem.

5.   Repeat Procedures 1–4 twice before rotating roles so that each group member performs each role twice.

Cooperative-Learning Activity
10.1 Number Magic With Rational Expressions

Group members: 2–4

Materials: calculator, grid paper, paper, and a pencil

Responsibilities: Each group will find the solution to a number problem by comparing rational expressions.

Preparation: The numbers 2 and 2 are two numbers whose sum equals their product. The sum is 4 and the product is 4. No matter what number you may choose there is always another number it can be added to or multiplied by to obtain the same result. In this activity, you will explore the rational expressions that make this happen.

Procedures:
1. Choose any number. Subtract one from the number. Write the reciprocal of this number. Add 1 to the reciprocal. Record the results on a separate piece of paper.

2. Find the sum of x and $1 + \dfrac{1}{x-1}$. Then find the product in the space provided.

3. As a group, describe what you notice. _____

_____

4. Choose at least 6 different numbers, including natural numbers, fractions, and decimals, and record them in a table similar to the one below that extends to the seventh number.

Number	x	$x-1$	$\dfrac{1}{x-1}$	$1 + \dfrac{1}{x-1}$	Sum	Product

5. The sum can be modeled by the rational equation $S = x + 1 + \dfrac{1}{x-1}$, where x is the number. As a group, graph the rational function represented by S on grid paper.

6. The product can be modeled by the rational equation $P = x\left(1 + \dfrac{1}{x-1}\right)$, where x is the number. On the same set of axes, graph the rational function that is represented by P.

7. As a group, compare the two graphs. Describe what you notice.

8. Describe how to find two numbers such that their sum and product are the

same. _____

Cooperative-Learning Activity
10.2 Scale Models and Inverse Variation

Group members: 4–6

Materials: calculator, ruler, paper, and a pencil

Responsibilities: The group will visualize the vastness of the solar system by representing planet sizes and distances from the sun on the same scale.

Preparation: Imagine our solar system. The Earth has a diameter of approximately 13,000 kilometers. It is about 800 million kilometers from the sun. Pluto is approximately 6 billion kilometers from the sun. In this activity, your group will place the planet models at the correctly scaled distances from each other by calculating the model distances and then imagining what the full model would look like if the planet models were spaced correctly.

In a scale model, model dimensions and real dimensions are proportional. The ratio of the model dimensions to the real dimensions must be the same constant. This constant is called the scale factor.

The table approximates the planet sizes and distances from the sun.

Planet	Real diameter (km)	Real distance from the Sun (km)
Sun	1,391,000	0.00
Mercury	4880	5.8×10^7
Venus	12,100	1.1×10^8
Earth	12,756	1.5×10^8
Mars	6791	2.3×10^8
Jupiter	143,200	7.8×10^8
Saturn	120,000	1.4×10^9
Uranus	51,800	2.9×10^9
Neptune	49,500	4.5×10^9
Pluto	3000	5.9×10^9

Procedures:

1. Create a table like the one above with planets in one row and real distance from the sun in one row, and then create two new rows, one for model diameter in centimeters and one for model distance from the sun in centimeters.

2. Use the scale factor 1.8×10^{-5} cm per km for the model diameter and 1.8×10^{-7} m per km for model distance. Note: These are the same scale factors. As a group, calculate the dimensions for this scale model and record the results in the table your group created.

3. Imagine modeling the sun with a basketball. Describe an object that could

   model Jupiter. _____

4. If these planet models were properly spaced from the model of the sun,

   where would Pluto be located? _____

Cooperative-Learning Activity
10.3 Creating Fractals

Group members: 4

Materials: tape, paper, scissors, ruler, paper, and a pencil

Roles: Tracer   traces a copy of the given squares and places letter *L* inside

Reducer   reduces the squares by 50% and makes copies of the result

Rotator   tapes and rotates the squares together

Drawer   draws the final result of the group's design

Preparation: A fractal is a complex shape that looks the same at all magnifications. In this activity, your group will create a fractal.

Procedures: 1.   The Tracer traces a copy of the square shown in the figure and places the letter *L* inside the square.

2.   By hand or with a copier, the Reducer reduces this square by 50% and makes two additional copies of the reduced square.

3.   The Rotator tapes the reduced squares together so that they share two adjacent sides following these conditions:

    a.   Keep one square stationary.
    b.   Rotate a second square 90° clockwise.
    c.   Rotate the third square 180° clockwise.

The result will look like the following design:

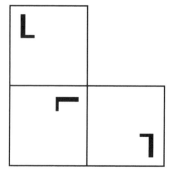

4.   Continuing the process, the Reducer reduces this design by 50% and makes two additional copies, as your group did in Procedure 3. The Drawer draws the final result on a separate piece of paper.

Cooperative-Learning Activity
10.4 Expressing Numbers in the Greatest Number of Ways

Group members: 4

Materials: calculator and a pencil

Responsibilities: Find the greatest number of ways to write a rational number.

Preparation: The ancient Egyptians used only fractions with 1 as the numerator. Such fractions are called unit fractions. They used sums of unit fractions and did not use the same unit fraction more than once in a sum. For example, to write $\frac{5}{6}$, they found the largest unit fraction contained in $\frac{5}{6}$, which is $\frac{1}{2}$, and added $\frac{1}{3}$. Thus, the fraction $\frac{5}{6}$ was written as $\frac{1}{2} + \frac{1}{3}$. The fraction $\frac{2}{5}$ was written as $\frac{1}{5} + \frac{1}{6} + \frac{1}{30}$.

Procedures: In this game, members of the group will compete with each other to find the greatest number of ways to write a rational number as the sum of two or more unit fractions.

1. The teacher will give the group a rational number.

2. Each member has two minutes to create as many sums as possible. The unit fractions used are recorded in the table below.

3. At the end of the two minutes, the group checks the results. For each correct sum, the student receives two points. For an incorrect sum, the student loses two points.

4. The team member with the highest score at the end of the round wins the round.

5. After each round is completed and a winner is determined, the game is repeated with a different rational number.

Name of student	Number	Unit fraction

Cooperative-Learning Activity
10.5 Nomographs as Early Calculators

Group members: 3

Materials: centimeter ruler, protractor, paper, and a pencil

Roles: **Drawer** uses a protractor to draw angles

Folder folds the paper according to procedures

Recorder labels and records group input and output

Preparation: Before the existence of calculators, students used nomograms, like the one shown, to perform calculations. Your group will construct a nomogram and use the nomogram to express relationships that are modeled by rational expressions in equations.

A nomograph is an alignment chart. Examine the nomograph.

The vertex of the angle is O. The sides of the angle are $\overrightarrow{OX}$ and $\overrightarrow{OY}$.

The measure of $\angle XOY$ is 120°. $\overrightarrow{OZ}$ bisects $\angle XOY$.

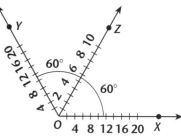

Procedures:
1. The Drawer uses a protractor to draw a 120° angle on a sheet of paper. Name the angle $\angle XOY$.

2. The Folder folds the paper so that $\overrightarrow{OX}$ falls on $\overrightarrow{OY}$. Lay the paper flat and call the fold line OZ. The Recorder records how $\overrightarrow{OZ}$ is related to $\angle XOY$. _____

3. The Drawer places 10 evenly spaced tick marks on $\overrightarrow{OX}$. Use the scale 0.5 cm = 2 units. The Recorder labels the tick marks with consecutive even integers from 2 to 20. Repeat the process for $\overrightarrow{OY}$.

4. The Folder places a ruler perpendicular to $\overrightarrow{OZ}$ and intersecting $\overrightarrow{OX}$ and $\overrightarrow{OY}$ at 2 and makes a tick mark at the point of intersection. Continue until 10 points are marked. The Recorder labels the tick marks with consecutive integers from 1 to 10.

5. As a group, use the nomograph to solve this problem:

   Rick can place tile on a floor in 12 hours. Elaine can place tile on a floor in 6 hours. How many hours will it take if they work together? _____

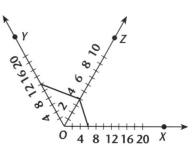

   a. Place a ruler so that it intersects $\overrightarrow{OX}$ at 12 and another rule intersecting $\overrightarrow{OY}$ at 6.
   b. The answer, 4 hours, is read at the point of intersection of the ruler and $\overrightarrow{OZ}$.

Cooperative-Learning Activity
10.6 Estimating the Number of Fish by Using Proportions

Group members: 3–6

Materials: lima beans, bowl, scoop, paper, and a pencil

Responsibilities: Use the tag-and-capture method to estimate how many fish are in a bowl.

Preparation: To determine the number of fish in a lake, a park ranger catches a number of fish, tags them, and throws them back into the lake. Later, a park ranger catches a second number of fish and counts the number tagged. A proportion is used to estimate the population of fish. In this activity, you will use the tag-and-capture method to estimate how many "fish" are in a bowl.

Procedures:

1. Fill the bowl with lima beans. The beans represent the fish. Create a table like the one below on a separate piece of paper to record the group's results.

Sample:	1	2	3	4	5	6
Tagged in sample: t						
Number in sample: n						
Total tagged: T						
Number in bowl: N						

2. One member takes a full scoop of fish from the bowl. As a group, count and mark each fish with a label. Then record this number in each cell of the row marked Total tagged. This will be your tagged sample. Put the fish back in the bowl.

3. For the samples, each member of the group shakes the bowl so that the fish are evenly distributed, takes out a full scoop of fish, and counts the number of tagged fish in the scoop. Return the fish and record your results. Repeat this until each member of your group has selected and counted a sample of fish.

4. Each member estimates the total number of fish in the bowl by using the proportion $\frac{T}{N} = \frac{t}{n}$ for their sample. Record your results. _____

5. Next count all of the fish in your bowl and calculate the percent error for each sample. Create a table like the one below on a piece of paper.

Sample	1	2	3	4	5	6
Percent error						

6. Notice the percent error. Would increasing the number of sample decrease the percent error? Would increasing the sample size decrease the percent error?

   _____

7. Describe how your group could improve the experiment in order to make an acceptable estimate of the total number of fish in the bowl.

   _____

Cooperative-Learning Activity
10.7 Proof in Algebra With Algebra Tiles

Group members: 2–4

Materials: algebra tiles and a pencil

Responsibilities: Use algebra tiles to verify identities.

Preparation: Examine the figure shown. If the figure is labeled as shown, the tiles illustrate the identity $(a - b)^2 = a^2 - 2ab + b^2$.

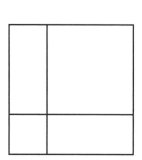
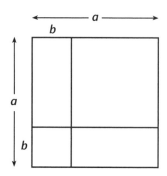

Procedures: 1. Label this figure to represent the following identity:

$$(a + b + c)^2 = a^2 + b^2 + c^2 + 2ab + 2ac + 2bc$$

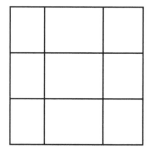

Each group member will arrange the algebra tiles to show one of the following identities. Each group member will draw his or her arrangement on a separate piece of paper. Then, as a group, examine each member's arrangement.

2. $4ab + (a - b)^2 = (a + b)^2$

3. $(a + b)^2 + (a - b)^2 = 2(a^2 + b^2)$

4. $(a + b)^2 = a^2 + 2ab + b^2$

ANSWERS

Cooperative-Learning Activity—Chapter 1

Lesson 1.1

Check students' work.

Lesson 1.2

Answers will vary depending on the room and scale selected.

Lesson 1.3

Answers will vary depending on the measurements taken.

Lesson 1.4

Check students' work.

Lesson 1.5

$x = 5$

Lesson 1.6

1. $|x + 1| < 4$
   $x < 3$ and $x > -5$

2. $|x + 7| < 2$
   $x > -9$ and $x < -5$

3. $|x - 8| \geq 1$
   $x \geq 9$ or $x \leq 7$

4. $|x + 5| = -3$
   no solution

5. $|x + 0| = 4$
   $x = 4$ and $x = -4$

6. $|x + 3| < 1$
   $-4 < x < -2$

7. $|x - 2| \geq 1$
   $x \geq 3$ or $x \leq 1$

8. $|-3 + x| > 0$
   $x \neq 3$

Cooperative-Learning Activity—Chapter 2

Lesson 2.1

3–5. Answers will vary. The class should note that the greater the slope, the steeper the incline.

Lesson 2.2

4. The data points fit a straight line.

5–6. Answers will vary.

Lesson 2.3

2.
Number of rubber bands	1	2	3	4
Number of intersections	0	1	3	6

4. Answers will vary. Number of intersections $= \frac{n(n - 1)}{2}$, where n is number of rubber bands

5–6. Answers will vary.

Lesson 2.4

1.
Left side	Right side	Equation
x	y	
$x - 1$	$y + 1$	$x - 1 = y + 1$
$x + 1$	$y - 1$	$x + 1 = 2(y - 1)$

2. 7 students on the left side and 5 students on the right side

Lesson 2.5

1. The solution for each system is $(-1, 2)$.

2. The solution is the same for each system.

3. The coefficients and constant are consecutive integers.

ANSWERS

4.

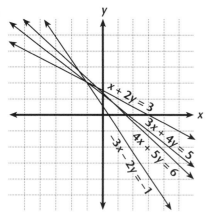

$x + 2y = 3$
$3x + 4y = 5$
$4x + 5y = 6$
$-3x - 2y = -1$

5. The lines intersect at $(-1, 2)$.

6. Answers will vary.

Lesson 2.6

3. There is one line segment less. It seems as though one line segment disappeared.

4. The line segment reappears.

7. There is one line segment less.

8. Eight of the 10 segments are divided into 2 segments. Then these segments form 9 line segments slightly longer than the original segments.

Lesson 2.7

1.

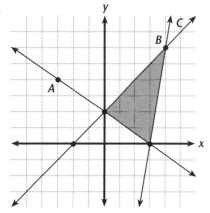

2–4. Answers will vary.

Lesson 2.8

1–5. Answers will vary.

6. The final sum is always 1089.

7. Answers will vary.

Cooperative-Learning Activity—Chapter 3

Lesson 3.1

Answers will vary.

Lesson 3.2

Answers will vary.

Lesson 3.3

Answers will vary.

Lesson 3.4

Answers will vary.

Lesson 3.5

1. a. $x = 54$ inches, $y = 70$ pounds

b. $x = 54$ inches, $y = 77$ pounds

c. $x = 54$ inches, $y = 63$ pounds

2. Students might notice that the heights are the same for males and females, but that the weights differ by 7 pounds, or about 10% of the weight of a person with average-size bones.

Cooperative-Learning Activity—Chapter 4

Lesson 4.1

Answers will vary.

The experimental probability is approximately 0.30.

Lesson 4.2

Answers will vary.

The probability of 7 or more correct is about 0.2, or 20%.

Lesson 4.3

Answers will vary. The class average is approximately between 160 and 200. The class mean is about 180.

The class average would be about 200.

Lesson 4.4

1–4. Answers will vary.

5. The first move can be made in 4 ways.

7. Answers will vary. The asterisk indicates the coin that was not turned over.

	T	T	T	T
First move	T*	H	H	H
Second move	T	H*	T	T
Third move	T	T	T*	H
Fourth move	H	H	H	H*

Lesson 4.5

There are 64 possible paths.

Lesson 4.6

2. 81, 41

3. $P(EVEN) = \frac{41}{81}$

$P(Odd) = \frac{40}{81}; \frac{1}{2}$

4. Game B is not a fair game. Probabilities for *ODD* winning or *EVEN* winning are not equally likely.
Game A is fair. Probabilities for Odd winning or Even winning are equally likely.

Lesson 4.7

1. 2 **2.** 6 **3.** $\frac{1}{2}$ **4.** $\frac{1}{4}$ **5.** $\frac{1}{4}$ **6.** $\frac{1}{2}$

7. $\frac{1}{4}$

Cooperative-Learning Activity—Chapter 5

Lesson 5.1

1. $y = x - 3$ **2.** $y = |x|$ **3.** $y = x^2$

4–5. Answers may vary.

Lesson 5.2

Answers may vary.

Lesson 5.3

Answers may vary.

Lesson 5.4

Answers may vary.

Lesson 5.5

Answers may vary.

Lesson 5.6

Answers may vary.

Cooperative-Learning Activity—Chapter 6

Lesson 6.1

1. Estimates will vary.

2. $2^0, 2^1, 2^2, \ldots, 2^{63}$
$2^3 = 8; 2^{15} = 32{,}768; 2^{63} = 9.22 \times 10^{18}$

3. The addend is one less than the sum.

ANSWERS

4. $2^{64} - 1$, or 18,446,744,073,709,551,615 quintillion

5. Answers will vary.

Lesson 6.2

6. 256 **7.** 144 **8.** 84 **9.** 1024

Lesson 6.3

1. The area of the folded sheet is $\frac{1}{2}$ of the area of the original sheet.

2. Check students' explanations.

3–4.

Number of folds	1	2	3	4
Area of folded paper	$\frac{1}{2}$	$\frac{1}{4}$	$\frac{1}{8}$	$\frac{1}{16}$

	5	6	7	n
	$\frac{1}{32}$	$\frac{1}{64}$	$\frac{1}{128}$	$\frac{1}{2^n}$

As the number of folds increases, the area decreases exponentially.

5.

Number of folds	1	2	3	4	5	n
Area of folded paper	$\frac{1}{3}$	$\frac{1}{9}$	$\frac{1}{27}$	$\frac{1}{81}$	$\frac{1}{243}$	$\frac{1}{3^n}$

6. This paper folding activity can be modeled by an exponential function.

Lesson 6.4

Answers will vary. Approximately 7 students can stand inside a square yard.

Lesson 6.5

4.

Number of markers	1	2	3	4	5
Number of moves	1	3	7	15	31

5. The graph is a transformation of $y = 2^x$.

6. 31 **7.** $y = 2^x - 1$

Lesson 6.6

The rate of reaction is affected by the temperature of the water. The higher the temperature, the faster the reaction. The increase in reaction rate is exponential.

Cooperative-Learning Activity—Chapter 7

Lesson 7.1

Number of subscriptions	Subscription price ($)	Total revenue ($)
100	27.50	2750.00
101	27.50 − (0.20 × 1)	2757.30
102	27.50 − (0.20 × 2)	2764.20
103	27.50 − (0.20 × 3)	2770.70
104	27.50 − (0.20 × 4)	2776.80
105	27.50 − (0.20 × 5)	2782.50
106	27.50 − (0.20 × 6)	2787.80
107	27.50 − (0.20 × 7)	2792.70
108	27.50 − (0.20 × 8)	2797.20
109	27.50 − (0.20 × 9)	2801.30
110	27.50 − (0.20 × 10)	2805.00
111	27.50 − (0.20 × 11)	2808.30
112	27.50 − (0.20 × 12)	2811.20
113	27.50 − (0.20 × 13)	2813.70
114	27.50 − (0.20 × 14)	2815.80
115	27.50 − (0.20 × 15)	2817.50
116	27.50 − (0.20 × 16)	2818.80
117	27.50 − (0.20 × 17)	2819.70
118	27.50 − (0.20 × 18)	2820.20
119	27.50 − (0.20 × 19)	2820.30
120	27.50 − (0.20 × 20)	2820.00
121	27.50 − (0.20 × 21)	2819.30

2. Answers will vary. **3.** 119; 119

4. $f(x) = (100 + x)(27.5 - 0.2x)$

Lesson 7.2

1. 13 **2.** 25 **3.** $n^2 + (n - 1)^2$ **4.** 761

5. Answers will vary.

Lesson 7.3

1. a. 6 **b.** 2 **c.** 8

ANSWERS

2–3.

Number of Squares

Size	1 row × 1 column
1 row × 3 column	3
2 rows × 3 columns	6
3 rows × 3 columns	9
4 rows × 3 columns	12

2 rows × 2 columns	3 rows × 3 column	Total
		3
2		8
4	1	14
6	2	20

4. Answers will vary. The number of squares are multiples of 3.

5. Answers will vary. The number of squares are multiples of 2.

6. 40

Lesson 7.4

1.

Geoboard size	1 × 1	2 × 2
Number of different line segment lengths	0	2

3 × 3	4 × 4	5 × 5	. . .	n × n
5	9	14		$\dfrac{(n + 2)(n - 1)}{2}$

2. Answers will vary. The number of different line segments increases by consecutive integers.

3. 20   **4.** $\dfrac{(n + 2)(n - 1)}{2}$

5. $\dfrac{n^2 + n - 2}{2}; \dfrac{((10)^2 + 10 - 2)}{2}; \dfrac{108}{2} = 54$

Lesson 7.5

4. Answers will vary.

5.

Number of markers of each color, n	1	2	3	4
Minimum number of moves	3	8	15	24

6. $n^2 + 2n$

Lesson 7.6

1–3.

Design	1 × 1	2 × 2
Number of vertices	6	16

3 × 3	4 × 4	. . .	n × n
30	48		$2n(n + 2)$

4. 16; 30, 5; 48, 6

5. $2n(n + 2)$; $18(11) = 198$

Lesson 7.7

3–4. Answers will vary.

Cooperative-Learning Activity—Chapter 8

Lesson 8.1

4.

Number of folds, x	0	1	2	3	4
Number of sections, f(x)	1	2	4	7	11

5. First differences: 1 2 3 4
The difference increases by 1.

6. Second differences: 1 1 1

7. The second differences are constant.

8. $f(5) = 16$; $f(6) = 22$
General formula: $f(x) = 0.5x^2 + 0.5x + 1$

Lesson 8.2

3–4. Answers will vary. Possible answer: (1.75, 6), (2, 13), (2.75, 24), (3, 29)

5. The curve is a transformation of $y = x^2$. The curve is narrower than $y = x^2$.

6. Answers will vary. Sample answer: (12, 452)

7. $f(x) = \pi x^2$

ANSWERS

Lesson 8.3

1. Answers will vary.

2. first figure: 6 y-tiles; second figure: 12 tiles

3. third figure: 20-tiles; answers will vary.

4. A rectangle is formed.

5. $(1 + 1)(1 + 1 + 1) = 6$

6. $(n + 1)(n + 2)$

Lesson 8.4

1. parabola

2–9. Answers will vary. Sample answer: Suppose that the length of the stream between x-intercepts is about 10 inches. A quadratic equation that models the stream of water is $y = 0.5x(10 - x)$. The maximum value is 12.5 inches, when x is 5 inches.

Lesson 8.5

1. Answers will vary.

2. one small rectangle: 1 rectangle; two small rectangles: 3 rectangles

3. 6 rectangles

4.

x	1	2	3	4	5
$f(x)$	1	3	6	10	15

5. The data fit a quadratic equation.

6. Answers will vary. The second differences are constant. The quadratic equation is $y = 0.5x^2 + 0.5x$.

7. $f(8) = 36$; $f(10) = 55$

Lesson 8.6

1–6. Answers will vary.

Sample disk measurements: dime = 18 mm, nickel = 21 mm, quarter = 24 mm, half-dollar = 29 mm, yogurt lid = 80 mm, tea cup = 88 mm, Frisbee = 261 mm, trash can = 515 mm, bike wheel = 660 mm

A quadratic function fits the data.

Cooperative-Learning Activity—Chapter 9

Lesson 9.1

1. The cooling effect of the wind is negligible at speeds less than or equal to 4 miles per hour.

2. $T_w = 1.60t - 55$, $v > 45$

3.

Windchill Temperature

Wind speed	Temperatures in degrees Fahrenheit			
	$-20°$	$-10°$	$-5°$	$0°$
10 mph	-45.5	-33.2	-27.1	-20.9
15 mph	-58.1	-44.7	-38.0	-31.3
20 mph	-67.0	-52.8	-45.7	-38.5
25 mph	-73.4	-58.6	-51.2	-43.8
30 mph	-78.2	-62.9	-55.3	-47.7
40 mph	-84.0	-68.3	-60.4	-52.5

	$5°$	$10°$	$20°$	$30°$	$40°$
10 mph	-14.8	-8.6	3.7	15.9	28.2
15 mph	-24.6	-17.9	-4.4	9.0	22.4
20 mph	-31.4	-24.3	-10.1	4.1	18.3
25 mph	-36.4	-29.0	-14.2	0.6	15.3
30 mph	-40.1	-32.5	-17.3	-2.1	13.2
40 mph	-44.7	-36.8	-21.0	-5.3	10.5

ANSWERS

Lesson 9.2

4. $\frac{5}{7} = 0.7142857143\ldots$

$\frac{12}{17} = 0.7058823529\ldots$

$\frac{29}{41} = 0.7073170732\ldots$

$\frac{70}{99} = 0.7070707070\ldots$

5. $\frac{1}{\sqrt{2}} = 0.7071067812\ldots$

The ratios get closer and closer to $\frac{1}{\sqrt{2}}$.

6. $\frac{2}{2} = 1$

$\frac{4}{6} = 0.66666\ldots$

$\frac{10}{14} = 0.7142857143\ldots$

$\frac{24}{34} = 0.7058823529\ldots$

$\frac{58}{82} = 0.7073170732\ldots$

$\frac{140}{198} = 0.7070707071\ldots$

The ratios approximate $\frac{1}{\sqrt{2}}$.

Lesson 9.3

1. $x = \dfrac{-1 + \sqrt{5}}{2}$ or $x = \dfrac{-1 - \sqrt{5}}{2}$

$x = 0.61803\ldots$

2. $\frac{15.5}{25} = 0.62$, which approximates the golden ratio

Lesson 9.4

1. $\overline{AB}, \overline{AC}, \overline{AD}, \overline{AE}, \overline{AF}, \overline{AG}, \overline{AH}, \overline{AI}$; the total number of line segments with A is 8.

2. $\overline{BC}, \overline{BD}, \overline{BE}, \overline{BF}, \overline{BG}, \overline{BH}, \overline{BI}$; the total number of line segments with B that do not include A is 7.

3. $\overline{CD}, \overline{CE}, \overline{CF}, \overline{CG}, \overline{CH}, \overline{CI}$; the total number of line segments with C that do not include A or B is 6.

4. $\overline{DE}, \overline{DF}, \overline{DG}, \overline{DH}, \overline{DI}, \overline{EF}, \overline{EG}, \overline{EH}, \overline{EI}, \overline{GH}, \overline{GI}, \overline{HI}$; the total number of line segments is 36.

5.

	A	B	C	D	E	F	G	H	I
A	0	1	2	1	$\sqrt{2}$	$\sqrt{5}$	2	$\sqrt{5}$	$2\sqrt{2}$
B	1	0	1	$\sqrt{2}$	1	$\sqrt{2}$	$\sqrt{5}$	2	$\sqrt{5}$
C	2	1	0	$\sqrt{5}$	$\sqrt{2}$	1	$2\sqrt{2}$	$\sqrt{5}$	2
D	1	$\sqrt{2}$	$\sqrt{5}$	0	1	2	1	$\sqrt{2}$	$\sqrt{5}$
E	$\sqrt{2}$	1	$\sqrt{2}$	1	0	1	$\sqrt{2}$	1	$\sqrt{2}$
F	$\sqrt{5}$	$\sqrt{2}$	1	2	1	0	$\sqrt{5}$	$\sqrt{2}$	1
G	2	$\sqrt{5}$	$2\sqrt{2}$	1	$\sqrt{2}$	$\sqrt{5}$	0	1	2
H	$\sqrt{5}$	2	$\sqrt{5}$	$\sqrt{2}$	1	$\sqrt{2}$	1	0	1
I	$2\sqrt{2}$	$\sqrt{5}$	2	$\sqrt{5}$	$\sqrt{2}$	1	2	1	0

6. Answers will vary. The table is symmetric along the main diagonal.

7. The distance from A to E is $\sqrt{2}$. The distance from A to I is $\sqrt{8}$. Since the distance from A to I is twice the distance from A to E, $\sqrt{8} = 2\sqrt{2}$.

8. 120

Lesson 9.5

2. a. 2
   b. 2
   c. 2
   d. 2
   e. 5

3. $\triangle BCF, \triangle BCG, \triangle BDF, \triangle BEG, \triangle BFG, \triangle BDG$; 7

4. $\triangle CDE, \triangle CDG, \triangle CEF, \triangle CEG, \triangle DEF, \triangle DEG, \triangle DFG, \triangle EFG$
The total number of triangles is 24.

5. The perimeter of $\triangle ACG$ is $4 + 2\sqrt{2}$. The area is 2.
The perimeter of $\triangle BDG$ is $2 + 2\sqrt{5}$. The area is 2.
The perimeter of $\triangle AEF$ is $4 + 2\sqrt{5}$. The area is 2.

Triangles with the same area do not necessarily have the same perimeter.

ANSWERS

6. The lengths of the sides of ΔFGB are 1, $\sqrt{2}$, and $\sqrt{5}$; $1^2 + (\sqrt{2})^2 = 3$, which is less than $(\sqrt{5})^2$; the triangle is scalene.

Lesson 9.6

4. Answers will vary. Measure each with a piece of string.

5. The length of a side of square *ABCD* is approximately 1.646. The area is approximately 2.7.

6. The area of circle *P* is π. The area of the square approximates the area of the circles.

Lesson 9.7

1.

Vertical height (ft)	Base of the ladder to the wall (ft)	Ladder length should be at least (ft)
12	4.4	12.6
14	5.1	14.9
16	5.8	17.0
20	7.3	21.3
24	8.7	25.5
28	10.2	29.8
32	11.6	34.1
34	12.4	36.2
38	13.8	40.4
44	16.0	46.8

2–3. Answers will vary

Lesson 9.8

Answers will vary.

Cooperative-Learning Activity—Chapter 10

Lesson 10.1

1. Answers will vary.

2. The sum is $\dfrac{x^2}{x-1}$. The product is $\dfrac{x^2}{x-1}$.

3. The sum equals the product.

4. Answers will vary.

5. The graph is a rational function.

6. The graph is a rational function.

7. The graphs are the same.

8. Choose any number. Subtract one from the number. Write the reciprocal of the number. Add 1 to the result. The sum and product will be equal.

Lesson 10.2

1.

	Real diameter (km)	Real distance from sun (km)
Sun	1,391,000	0.00
Mercury	4880	5.8×10^7
Venus	12,100	1.1×10^8
Earth	12,756	1.5×10^8
Mars	6791	2.3×10^8
Jupiter	143,200	7.8×10^8
Saturn	120,000	1.4×10^9
Uranus	51,800	2.9×10^9
Neptune	49,500	4.5×10^9
Pluto	3000	5.9×10^9

	Model diameter (cm)	Model distance from sun (m)
Sun	25.04	0.0
Mercury	0.09	10.4
Venus	0.22	19.8
Earth	0.23	27.0
Mars	0.12	41.4
Jupiter	2.58	140.4
Saturn	2.16	252.0
Uranus	0.93	522.0
Neptune	0.89	810.0
Pluto	0.05	1062.0

2. If the basketball were 25 cm in diameter, Jupiter would be an object that is an inch in diameter.

3. Pluto would be over 1 km away from the sun modeled by a basketball.

ANSWERS

Lesson 10.3

4.

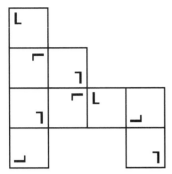

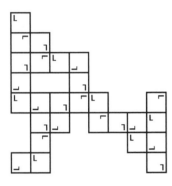

Lesson 10.4

Answers will vary.

Lesson 10.5

1–5. Answers will vary.

Lesson 10.6

3–5. Answers will vary.

6. Increasing the sample size should decrease the percent error.

7. Answers will vary. Sample answer: Increase the number of samples.

Lesson 10.7

1.

	a	b	c
a	a^2	ab	ac
b	ab	b^2	bc
c	ac	bc	c^2

2.

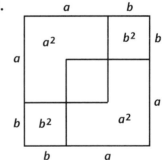

3.

4.